A
PRACTICAL GUIDE
TO
COMPUTER METHODS
FOR
ENGINEERS

A
PRACTICAL GUIDE
TO
COMPUTER METHODS
FOR
ENGINEERS

TERRY E. SHOUP

University of Houston

Prentice-Hall, Inc., Englewood Cliffs, N.J. 07632

Library of Congress Cataloging in Publication Data

Shoup, Terry E., 1944-
    A practical guide to computer methods for
engineers.

    Includes bibliographical references and index.
    1.   Engineering--Data processing.
2.   Engineering--Mathematical models. I. Title.
TA345.S56   620'.00285'4   78-10467
ISBN 0-13-690651-6

©1979 by Prentice-Hall, Inc., Englewood Cliffs, N.J. 07632

Printed in the United States of America

10 9 8 7 6 5 4 3 2 1

PRENTICE-HALL INTERNATIONAL, INC., London

PRENTICE-HALL OF AUSTRALIA PTY. LIMITED, Sydney

PRENTICE-HALL OF CANADA, LTD., Toronto

PRENTICE-HALL OF INDIA PRIVATE LIMITED, New Delhi

PRENTICE-HALL OF JAPAN, INC., Tokyo

PRENTICE-HALL OF SOUTHEAST ASIA PTE. LTD., Singapore

WHITEHALL BOOKS LIMITED, Wellington, New Zealand

To my parents Betty and Dale Shoup

CONTENTS

Misc

PREFACE

The present revolution in size, cost and capability of modern computational equipment has led to the application of computer methodology to a diversity of technological areas. With the increased emphasis on computer applications in engineering problem solving, more and more engineers are relying on the computer as a tool to assist them in their work. In most practical situations, the engineering programmer will build his or her program from the building blocks of available algorithms. This programming approach is based on sound judgment because it saves time and allows the engineer to exploit the computational talents of his or her computer science and mathematics colleagues. Although most practicing engineers and engineering students have a good working knowledge of computer programming, most are not well acquainted with the characteristics and limitations of the computer methods available to solve their problems. This book is aimed specifically at meeting this need--namely, to provide practical insight into selection of the best available algorithm to perform a given computational task. Unlike most texts in numerical methods, this book is written by an engineer specifically for use by engineers. The philosophy of presentation in this text begins with a clear, basic presentation of the fundamental algorithms. This is done with narrative, figures, logic flow diagrams, and example applications. Once the various algorithms are understood, they are compared and discussed with a view toward their unique advantages or limitations in a particular engineering situation. Whenever possible, available software is listed along with guidelines for practical application. A brief appendix of available computer software for engineering applications is presented at the end of the text.

This book has been developed and refined as a result of teaching and research experience gained at Rutgers University and at the University of Houston. During its evolution this material has successfully been presented to engineers and engineering students from a diversity of educational and experience backgrounds. The order of presentation for the topics in this text is one in which material in the early part of the book provides the necessary background for later chapters. The level of treatment presented is satisfactory for upper level undergraduates, graduate students, and practicing engineers. The spectrum of topics discussed in this text treats those problems most frequently encountered in engineering problem solving. References are provided at the end of each chapter for the reader who would like to pursue the topics in more depth.

The author would like to thank all of those who contributed encouragement and creative suggestions for the construction of this text and the approach it provides. Special thanks for encouragement and support are due to Dr. R. H. Page and Dr. L. C. Witte. Special appreciation must be expressed to my faculty colleagues Dr. R. Bannerot and Dr. M. Milleur, who contributed helpful suggestions and information. The author is grateful to S. Bass, L. Bilowich, J. Herrera, L. Sanchez, R. Strong, T. Wu and P. Young for reviewing the manuscript. Special thanks are due to R. Sodhi, K. Somkearti, and T. Zimmerman for their contributions to the "end of chapter" problems. Finally, I want to express my gratitude to my family for their patience and encouragement during the long hours spent on this project.

Terry E. Shoup

A
PRACTICAL GUIDE
TO
COMPUTER METHODS
FOR
ENGINEERS

1 INTRODUCTION

Because of its convenience and versatility, the digital computer is an extremely important tool in engineering problem solving. [Photo courtesy of Tektronix, Inc.]

1.1 Introduction to
Computer-Aided Problem Solving

Within the spectrum of technological functions, perhaps no activity holds more excitement than that of engineering problem solving. Indeed, a large portion of the iterative nature of engineering problem solving is embodied in the creative nature of design. Over the years, engineers have sought ways to speed up the mundane aspects of their function in order to free their time for the more creative aspects of their calling. Because of its ability to perform routine mathematical computations with extreme speed and accuracy, the digital computer is a logical tool for the engineer to exploit. As is true for the application of any new technological tool or process, the new freedom implies new responsibility. Whenever potential for high speed exists, so also does the potential for inefficiency and misapplication. Many of the difficulties associated with the use of the computer in engineering come not from the computer itself, but rather from the need for selection and application of the proper algorithm to solve a particular problem. By taking a few minutes before solving a problem to select the proper method, a designer can often save hours of wasted effort. It is the purpose of this book to provide insight into the proper application of the computer to perform a given design task. In this way the engineer can minimize frustration and maximize achievement. Unlike traditional courses in numerical methods, this course focuses attention on providing a variety of methods to solve a particular task. Particular emphasis is placed on an explanation of the intrinsic pitfalls and advantages that characterize a particular algorithm.

1.2 The Computer as a Tool
in the Design Environment

In order to understand when the computer can and cannot be applied in the engineering process, it is well for us to look briefly at the structure of design.

Engineering is a profession that employs both science and art to make a process or product for the benefit of mankind. It is different from any other field because it attempts to go from theory into practice rather than merely observing the phenomena of that science or art. The engineering process usually begins with the recognition that a need exists. It is the job of the engineer to create a usable solution to satisfy that need and to communicate that solution in sufficient detail so that the product or process can be produced. The sequence of events in the design process is shown in Figure 1-1. Here we see that the designer begins by under-

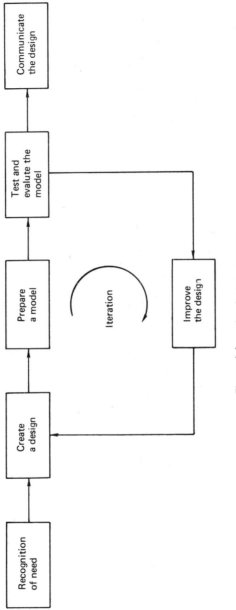

Figure 1-1 The design process.

3

taking a deliberate, planned exploration for possible solutions. Next, he or she assembles sufficient information to create a model suitable for the evaluation and testing of the utility of the solution. This phase of the activity is an economic convenience because to test the solution itself is nearly always too expensive, too time consuming, and too wasteful of materials and energy. Very often the idealized model used will be an analytical one or the combination of a simple prototype coupled with an analytical representation. Once the model has been prepared, the designer proceeds to test and evaluate its performance to obtain some measure of its ability to satisfy the need. Although the first attempts to create a solution to a design problem will usually produce a poor design, these first efforts will usually provide insight into where improvement can be made. Using these initial discoveries, the designer repeats the model building and evaluation cycle until he or she is satisfied that the best possible solution has been achieved. At this point the designer will communicate the design to the construction phase of the engineering process.

The various steps in the design process often consume different amounts of time. The boundaries between these steps are often difficult to observe, and frequently two or more steps will be combined. The total process is very iterative and parts can involve large amounts of computational effort. For this reason the computer can be an extremely effective aid if it is applied in an efficient way. The key to its application is to utilize it for mundane, repetitive tasks involving routine manipulations of data rather than for tasks that require the creative manipulation of abstract concepts. Thus, those tasks early in the design process are poorly suited to computer assistance whereas those at the end are best suited. An illustration of this situation is presented in Figure 1-2.

	ASPECT	COMPUTER POTENTIAL
	Recognition of Need	Desirable — but few applications exist
	Create a Solution	Limited — computer-augmented creativity
	Prepare a Model	Some — special languages and applications packages
	Test and Evaluate the Model	Yes — through numerical methods
	Improve the Design	Yes — optimization
	Communicate the Design	Yes — computer graphics, numerically controlled manufacturing

(left axis: Most creative ↑ ... Most potential for computer assistance ↓)

Figure 1-2 The potential for computer assistance in design.

Because of the wide scope of application shown in this figure, the computer should no longer be regarded as a supplementary tool in design. Rather, it has rapidly become a vital part of the design methodology utilized in the modern industrial environment.

1.3 Considerations in the Preparation of Engineering Programs

The types of calculations necessary for the design process are diverse in scope. Both the nature and the quantity of these calculations are continuously changing. Some computational procedures are sufficiently simple so as not to justify the use of the computer, while others involve such detail that they would be impossible without computer assistance. Those design tasks that are suitable for computer assistance may be classified as follows:

1. computations similar to hand calculations that must be done many times;

2. computations that are extensions of hand calculations but are too involved to be practical by hand for reasons of accuracy and time; and

3. manipulation of data for purposes of visualization, manufacture, or documentation.

1.4 Computational Equipment for Engineering Problem Solving

The array of computer equipment available for problem solving in the engineering environment is presented in Figure 1-3. The spectrum of computational hardware is presently undergoing a dramatic revolution in terms of size, cost, capability, and utility.

Due to recent advances in storage capacity and speed of access, the modern digital computer system is capable of handling extremely complex and large scale engineering problems. The software for large scale computer systems tends to be universal; thus programs written for one machine can generally be adapted for use on other systems of similar size. Through the medium of time-sharing terminals connected via telephone lines, the versatility and utility of the large-scale digital computer can be available at the desk of every engineering problem solver.

If large size is not required, and if the engineer wishes to avoid the need to share the computer with other users, he or she may prefer to utilize a minicomputer dedicated to specific problem-solving needs. These devices operate in much the same way as their larger counterparts except that they have smaller internal storage capacity and they allow consider-

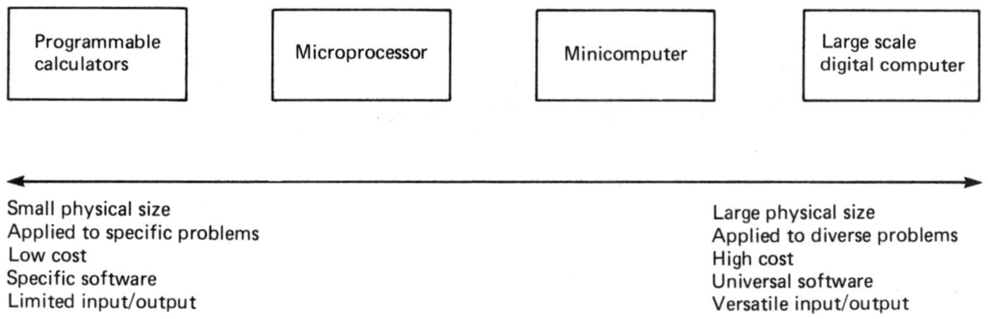

Programmable calculators | Microprocessor | Minicomputer | Large scale digital computer

Small physical size
Applied to specific problems
Low cost
Specific software
Limited input/output

Large physical size
Applied to diverse problems
High cost
Universal software
Versatile input/output

Figure 1-3 The spectrum of computational equipment for engineering problem solving.

able "hands-on" operation. This means that the engineering programmer can exercise greater personal control over the data management of problems.

If the computer in an engineering situation is to function as an integral part of an engineering system and is to be permanently dedicated to this task, a microprocessor may provide an efficient hardware solution. With such a device, the inputs frequently take the form of digitized measurements of the physical process being controlled, and the outputs typically take the form of signals or impulses that control the operation of other parts of the total system. Because of their low cost and the large domain of their practical applications, microprocessors have found their way into many aspects of our daily existence.

Recent advances in the field of programmable pocket calculators have made these miniature miracles capable of computational tasks surpassing those of larger computer systems of only a few years ago. Unlike most large scale computers, these personalized computers tend to use programming languages that are unique for the particular device to which they are applied. Programmable pocket calculators are well suited for problems requiring a small number of input values and a small number of output values. In most cases the numerical output must be displayed one number at a time. Even with their minor limitations, the price and convenience of these devices makes them readily available to every engineer. They continue to gain widespread popularity.

The actual selection of a particular type of computational device depends on the specific needs of the engineer and on the computational complexity of a particular engineering task. Since the array of available devices is a spectrum, it is not uncommon to find that several different types of devices are capable of performing a given computational task. In these situations the selection process would, of course, be based on cost, equipment availability, and personal preference.

1.5 Categories of Problems in Computer-Aided Engineering

Because of the nature of engineering phenomena, certain types of mathematical analysis problems occur again and again. Among these are:

1. the solution of algebraic equations;
2. the solution of eigenvalue problems;
3. the solution of ordinary differential equations;
4. the solution of partial differential equations;
5. the solution of optimization problems; and
6. the manipulation of numerical data.

Each of these areas is discussed in this book, and the fundamental concepts and terminology associated with the basic solution techniques for each of these areas are presented. In many cases the computer algorithms for solving engineering problems can be implemented on a variety of different computational devices. Wherever possible, familiar software is dicussed with a view toward making the best practical selection for a particular engineering problem situation.

References

1. Furman, T. T. The Use of Computers in Engineering Design. New York: Van Nostrand Reinhold Co., 1970.

2. Gibson, J. E. Introduction to Engineering Design. New York: Holt, Rinehart & Winston, 1968.

3. Klingman, E. E. Microprocessor Systems Design. Englewood Cliffs, N.J.: Prentice-Hall, Inc., 1977.

4. Mischke, C. R. An Introduction to Computer-Aided Design. Englewood Cliffs, N.J.: Prentice-Hall, Inc., 1968.

5. Smith, J. M. Scientific Analysis on the Pocket Calculator. New York: John Wiley & Sons, 1975.

6. Woodson, T. T. Introduction to Engineering Design. New York: McGraw-Hill Book Co., 1966.

2 COMPUTER SOLUTION OF ALGEBRAIC EQUATIONS

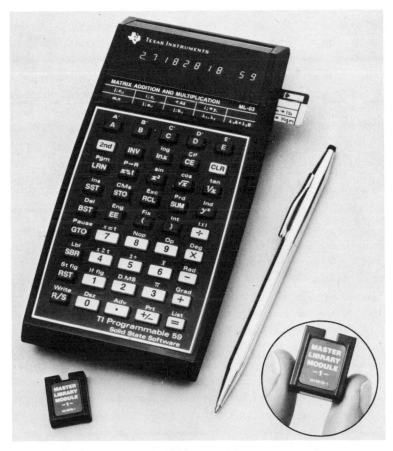

Because of its compactness and personal convenience, the programmable calculator is rapidly gaining acceptance as an indispensable engineering computational tool. [Photo courtesy of Texas Instruments, Incorporated.]

2.1   Introduction

The need to solve algebraic problems is a frequent situation in engineering design and analysis.   Problems of this type may occur either as complete problems by themselves or as contributing parts to more complex procedures.   In either case, the speed and efficiency with which the solution can be extracted will greatly influence the engineering utility of the computational process.   The selection of an appropriate algorithm for the solution of an algebraic system depends on the class of problem being considered.   Algebraic problems may be classified first according to the number of equations being solved and then according to the type and number of answers expected.   A classification diagram for algebraic problems is shown in Figure 2-1.   For the case of only one equation, the problem will be called linear, polynomial, or transcendental depending on whether it has one solution, "n" solutions, or an undetermined number of solutions.   In the case of multiple equations, problems will be called linear or nonlinear depending on the mathematical nature of the equations involved.

The solution of a single, linear equation for one unknown is a sufficiently straightforward process so that it will not be discussed in this chapter.   It is the purpose of this chapter to discuss various alternate methods for the solution of the other four remaining types of algebraic systems.

2.2   Roots of a Single Nonlinear Equation

Two common classifications of nonlinear algebraic equations are the transcendental equation and the polynomial equation.   Even though the techniques for the solution of these two categories are often the same, in this chapter they are treated separately because polynomials possess special solution properties.   We will consider nonpolynomial equations in this section and polynomial equations in the next section.

Nonlinear equations involving trigonometric relationships such as $\sin(x)$, $\cos(x)$, $\tan(x)$, or involving special functions such as $\log(x)$, e^x, etc., are called transcendental.   Solution methods for nonlinear equations of this type are either direct or indirect.   The direct methods find a solution by means of a set of formulas applied in a nonrepetitive fashion. The result is always an exact solution.   A familiar example of a direct method is the quadratic formula technique for second order polynomial equations.   An indirect method, on the other hand, provides a process for

10

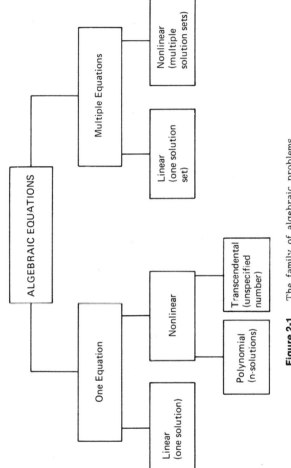

Figure 2-1 The family of algebraic problems.

11

achieving a solution as a result of the repeated application of an algorithm. The result is always approximate, although any desired degree of accuracy can usually be achieved. Iterative methods are best suited for computer implementation and are discussed in detail in this chapter. In each of the methods presented, it is assumed that the problem to be solved is that of finding the real roots (zeros) of the equation $f(x) = 0$. Although it is possible for nonpolynomial equations to have complex roots, the determination of complex roots is generally considered only for polynomial problems.

Binary Search Method

The logic flow diagram for the binary search method is presented in Figure 2-3 and proceeds as follows. First, the function is evaluated at equally spaced intervals of x until two successive function values $f(x_n)$ and $f(x_{n+1})$ are of opposite signs. (A change in sign indicates the existence of a root if the function is continuous.) On the range from x_n to x_{n+1}, the midpoint value is calculated using the formula:

$$x_{mid} = \frac{(x_{n+1} + x_n)}{2}$$

and the function value $f(x_{mid})$ is found. If the sign of $f(x_{mid})$ agrees with $f(x_n)$, it is used to replace $f(x_n)$; if not, it agrees with $f(x_{n+1})$ and replaces this value. The interval of uncertainty in which the root must lie is thus reduced. If the value of $f(x_{mid})$ is small enough, the process is terminated, otherwise, the process is repeated. Figure 2-2 illustrates this

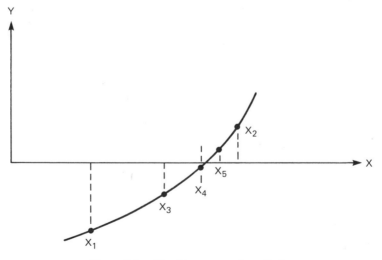

Figure 2-2   The binary search method.

12

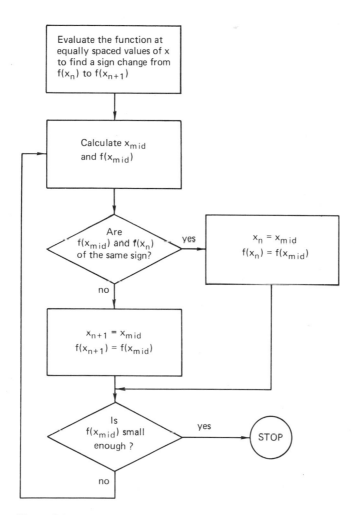

Figure 2-3 Logic flow diagram for the binary search method.

procedure graphically. Although this method is not efficient in computational effort, it does give an improved approximation to the actual root with increasing number of evaluations. Once the first interval of uncertainty is identified, the interval of uncertainty can be reduced by a factor of 2^{-N} for "N" iterations.

False Position Method

The method of false position is based on a linear interpolation between two values of the function that have opposite signs. This method often converges to a root more quickly than the binary search method. The logic flow diagram for this method is presented in Figure 2-4 and proceeds

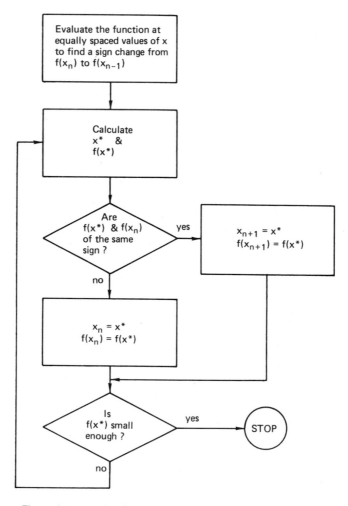

Figure 2-4 Logic flow diagram for the method of false position.

as follows. First the function is evaluated at equally spaced intervals of x until two successive function values $f(x_n)$ and $f(x_{n+1})$ are found to have opposite signs. A line passing through these two points will have a root at

$$x^* = x_n - f(x_n) \left(\frac{x_{n+1} - x_n}{f(x_{n+1}) - f(x_n)} \right).$$

This value is used to find $f(x^*)$, which is in turn used to compare with $f(x_n)$ and $f(x_{n+1})$ to replace the one with similar sign. If $f(x^*)$ is not close enough to zero, the calculation process is repeated until the desired degree of convergence is achieved. Figure 2-5 illustrates this process graphically.

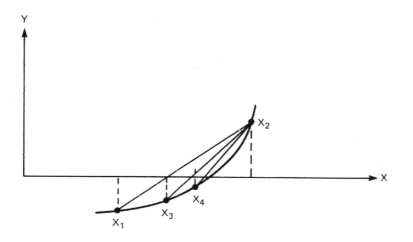

Figure 2-5   The false position method.

Newton's Method

Newton's method of iteration is the most widely used of all iterative techniques. Its popularity is due to the fact that, unlike the previous methods, it does not require the user to search for two function values of different sign in order to bracket the root. Rather than using an interpolation scheme based on two function values, this method uses extrapolation based on a line that is tangent to the curve at a point. The logic flow diagram for this method is shown in Figure 2-6. The method is developed from a Taylor's expansion of the form:

$$f(x_n + h) = f(x_n) + hf'(x_n) + \frac{h^2}{2} f''(x_n) + \cdots$$

The h^2 and higher order terms are dropped, and $x_n + h = x_{n+1}$ is used. It is assumed that the step from x_n to x_{n+1} moves the function value close

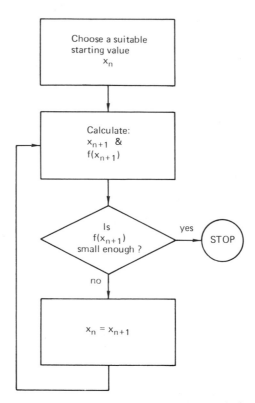

Figure 2-6   Logic flow diagram for Newton's method.

to a root so that $f(x_n + h) = 0$.   Then:

$$x_{n+1} = x_n - \frac{f(x_n)}{f'(x_n)} \ .$$

The value x_{n+1} is equivalent to the point where the curve tangent at x_n passes through the x-axis.   Since the curve $f(x)$ is likely not a line, the functional value $f(x_{n+1})$ is likely not exactly zero.   For this reason the process is repeated using $x_n = x_{n+1}$ as a new base point.   When the value of $f(x_{n+1})$ is sufficiently small, the process is terminated.   Figure 2-7 illustrates Newton's method graphically.   Clearly, the choice of location for the starting point will greatly influence the speed of convergence.   If the slope of the curve $f'(x)$ goes to zero in the iterative process, the method has difficulty.   In addition, it can be shown that if $f''(x)$ goes to infinity, the method will also fail to perform properly.   Since the conditions for a multiple root are $f(x) = 0$ and $f'(x) = 0$, Newton's method will not converge

16

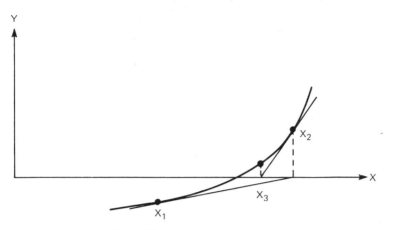

Figure 2-7 Newton's method.

for this case. It should be noted that an alternative procedure sometimes used for checking convergence is a comparison of x_n and x_{n+1}.

Secant Method

One of the disadvantages of Newton's method is that it requires the user to take a derivative of the function $f(x)$. In the event that it is inconvenient to find this derivative, an approximation may be used. This alternative is the basis for the secant method. If the derivative $f'(x_n)$ in the Newton method formula:

$$x_{n+1} = x_n - \frac{f(x_n)}{f'(x_n)}$$

is replaced by means of two successive functional approximations in the formula:

$$\text{Slope}(x_n) = \frac{f(x_n) - f(x_{n-1})}{x_n - x_{n-1}}$$

the iteration formula becomes:

$$x_{n+1} = x_n - \frac{f(x_n)}{\text{Slope}(x_n)}.$$

The logic flow diagram for this method is the same as for Newton's method except that the iteration formula is slightly different. This method actually seeks the root by a combination of interpolation and extrapolation. Whenever it is operating in the interpolation mode, the method is equivalent to the method of false position. As with the Newton method, this technique may be terminated when consecutive values of x agree to within some acceptable error or when the function value $f(x)$ is acceptably close

17

to zero. The secant method has the same convergence difficulties at a multiple root as does the method of Newton iteration.

Direct Substitution Method

The direct substitution method is a straightforward technique that can be used if the function $f(x) = 0$ can be manipulated into the form:

$$x = g(x).$$

Using this form, an iterative formula,

$$x_{n+1} = g(x_n)$$

can be formulated. The method would proceed in the pattern indicated in Figure 2-8. Because of its basic simplicity, this method may appear to be attractive; however, the method does have convergence pitfalls. For this reason any program using this algorithm should contain suitable checks to terminate the iterative process if it fails to converge.

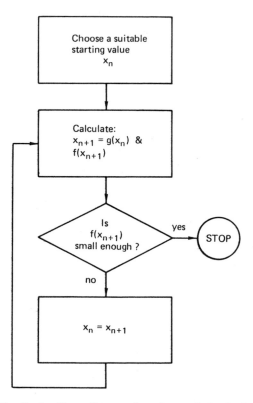

Figure 2-8   Logic flow diagram for the method of direct substitution.

18

Example 2-1

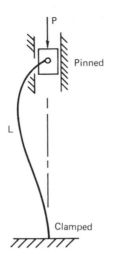

Suppose that it is desired to compute the buckling load for the clamped-pinned column shown in the figure above. The algebraic relationship describing this critical load will be:

$$\tan\sqrt{\frac{P}{EI}}\,L = \sqrt{\frac{P}{EI}}\,L$$

where:
P is the critical load for buckling,
EI is the flexural rigidity, and
L is the length of the column.

For purposes of simplification, the substitution

$$\sqrt{\frac{P}{EI}}\,L = x$$

can be made. Thus the transcendental equation to be solved is:

$$x - \tan(x) = 0$$

The method of Newton iteration will be used. In order to obtain an initial guess for the solution, a simple sketch of tan(x) vs x can be used to reveal that the first nonzero root lies between π and $3\pi/2$. Thus an initial guess of x = 4.5 will be used. A computer program that solves this problem is:

```
C       ***********************************************
C       *    THIS PROGRAM FINDS A SOLUTION TO THE      *
C       *    TRANSCENDENTAL EQUATION-                  *
C       *                                              *
C       *      F(X) = X - TAN(X) = 0                   *
C       *                                              *
```

```
C     *    BY THE METHOD OF NEWTON ITERATION.           *
C     *    THE PROCESS TERMINATES WHEN THE VALUE OF     *
C     *    F IS LESS THAN 0.00001 OR IF THE NUMBER      *
C     *    OF ITERATIONS EXCEEDS 30.                    *
C     *                      T. E. SHOUP  10/25/77      *
C     **********************************************************
C
C         CHOOSE A STARTING VALUE
          X = 4.5
          WRITE(6,100)
    100   FORMAT(1X,20('-'))
          WRITE(6,101)
    101   FORMAT(' ITERATION',2X,'    X')
          WRITE(6,102)
    102   FORMAT('  NUMBER  ',2X,'   VALUE')
          WRITE(6,100)
          WRITE(6,104) X
    104   FORMAT(' START',5X,F10.4)
          DO 1 I=1,30
C
C         CALCULATE THE FUNCTION VALUE
          F = X - SIN(X)/COS(X)
C
C         CALCULATE THE DERIVATIVE OF THE FUNCTION AT X
          DF = 1. - 1./(COS(X)**2)
          X = X - F/DF
          WRITE(6,103) I,X
      1   IF(ABS(F).LE.0.00001) GO TO 2
    103   FORMAT(1X,I5,5X,F10.4)
      2   WRITE(6,100)
          STOP
          END
```

The output of this program is:

```
--------------------
ITERATION      X
  NUMBER      VALUE
--------------------
START         4.5000
  1           4.4936
  2           4.4934
  3           4.4934
--------------------
```

Thus the method converges very rapidly from the initial value of 4.5. It is interesting to note, however, that if an initial guess of x = 4.0 or x = 5.0 is made, this method does not converge.

2.3   Solution of Polynomial Equations

Algebraic equations involving only the sum of integer powers of x are called polynomial equations. Their general form will be:

$$a_n x^n + a_{n-1} x^{n-1} + \ldots \ldots + a_1 x^1 + a_0 = 0$$

Properties of Polynomials
Certain special properties about polynomial equations are useful in

determining the nature of the solutions to these equations. These are as follows:

1. A polynomial of order "n" will have n-roots. These roots may be real or complex.

2. If all of the a_i coefficients are real, then all complex roots will appear in complex conjugate pairs.

3. The number of positive real roots is equal to or less than (by an integer), the number of sign changes in the a_i coefficients.

4. The number of negative real roots is equal to or less than (by an integer), the number of sign changes in the a_i coefficients if x is replaced by -x.

Although direct methods are available for the solution of second- and third-order polynomial equations, for higher-order equations indirect methods must be used. In the strict mathematical sense, once a root of a polynomial equation has been found by an indirect method, the root may be used to reduce the order of the polynomial by one by dividing by the linear factor (x - root). The result will be a polynomial or order n - 1. Although this procedure might seem attractive, it is not recommended because slight errors in the value of the initial root can cause the accumulation of substantial errors in the coefficients of the reduced polynomial. It should be noted that this procedure often does provide a means for selecting a reasonable guess for other roots once a few initial roots are already known.

Extensions of Previous Methods

The solution algorithms for transcendental equations presented in the previous section can be used to find both real and complex roots of polynomials if the user is willing to utilize complex arithmetic. The following simple example illustrates this fact.

Example 2-2

Suppose that it is desired to find the roots of the polynomial:

$$x^4 - x^3 - 4x^2 + 34x - 120 = 0$$

The method of Newton iteration using complex arithmetic will be used. An initial guess of x = 4. + 4i is used in the following computer program.

```
C      *******************************************************
C      *     THIS PROGRAM FINDS A SOLUTION TO THE           *
C      *        POLYNOMIAL EQUATION-                        *
C      *                                                    *
```

```
C      *      F(X) = X**4-X**3-4.*X**2+34.*X-120.         *
C      *                                                 *
C      *      BY THE METHOD OF NEWTON ITERATION.         *
C      *      THE PROCESS TERMINATES IF THE MAGNITUDE OF *
C      *      F IS LESS THAN 0.00001 OR IF THE NUMBER    *
C      *      OF ITERATIONS EXCEEDS 30.                  *
C      *                          T. E. SHOUP  10/25/77  *
C      ***************************************************
       COMPLEX X,F,DF
C
C      CHOOSE A STARTING VALUE
       X = (4.,4.)
       WRITE(6,100)
  100  FORMAT(1X,35('-'))
       WRITE(6,101)
  101  FORMAT(' ITERATION    X(REAL)       X(IMAG)')
       WRITE(6,102)
  102  FORMAT('   NUMBER')
       WRITE(6,100)
       WRITE(6,104) X
  104  FORMAT(' START',5X,F10.5,5X,F10.5)
       DO 1 I=1,30
C
C      CALCULATE THE FUNCTION VALUE
       F = X**4-X**3-4.*X**2+34.*X-120.
C
C      CALCULATE THE DERIVATIVE OF THE FUNCTION AT X
       DF = 4.*X**3-3.*X**2-8.*X+34.
       X = X - F/DF
       WRITE(6,103) I,X
       TEST = SQRT(REAL(F)**2+AIMAG(F)**2)
    1  IF(TEST.LE.0.00001) GO TO 2
  103  FORMAT(1X,I5,5X,F10.5,5X,F10.5)
    2  WRITE(6,100)
       STOP
       END
```

The output of this program is:

ITERATION NUMBER	X(REAL)	X(IMAG)
START	4.00000	4.00000
1	3.01187	3.02070
2	2.10319	2.29025
3	0.74130	1.88393
4	2.11373	3.58713
5	1.52897	2.95183
6	1.01802	2.85950
7	0.99751	3.01033
8	0.99998	3.00005
9	1.00000	3.00000
10	1.00000	3.00000

The iterative process converges to the root x = 1. + 3i. Since it is known in advance that one of the other roots must be a complex conjugate of this one, only two roots remain to be found. These roots may be estimated by applying the quadratic formula to the reduced polynomial and the iterative method may then be applied to the complete polynomial in order to refine their values. Even if this approach is not used, a different arbitrary initial guess will often lead to a different root. For example, if the guess x = 2 + 2i is used in this program, the root found will be x = 3. + 0i. The output of the computer program for this case is:

```
----------------------------------
ITERATION      X(REAL)      X(IMAG)
  NUMBER
----------------------------------
START          2.00000      2.00000
    1          0.11293      1.45022
    2          2.68064      1.81840
    3          1.63350      0.83697
    4          4.41050     -1.76305
    5          3.38850     -1.19750
    6          2.72681     -0.52745
    7          2.87133      0.12386
    8          2.99991     -0.01506
    9          2.99990      0.00000
   10          3.00000     -0.00000
   11          3.00000     -0.00000
----------------------------------
```

There are, of course, a number of standard polynomial root-solving routines available to the engineering programmer who chooses to use one of these. One such routine is the subroutine POLRT from the IBM Scientific Subroutine Package. This routine uses the method of Newton iteration to find all "n" roots of a polynomial of order "n." A computer program that solves this example problem using POLRT is:

```
C    ******************************************************
C    *      THIS PROGRAM FINDS A SOLUTION TO THE          *
C    *      POLYNOMIAL EQUATION-                          *
C    *                                                    *
C    *      F(X) = X**4-X**3-4.*X**2+34.*X-120.           *
C    *                                                    *
C    *      USING SUBROUTINE POLRT FROM THE               *
C    *      IBM SCIENTIFIC SUBROUTINE PACKAGE.            *
C    *                       T. E. SHOUP   10/25/77       *
C    ******************************************************
     DIMENSION XCOF(5),COF(5),ROOTR(4),ROOTI(4)
C
C    LOAD THE COEFFICIENT VECTOR
     XCOF(1) = -120.
     XCOF(2) = 34.
     XCOF(3) = -4.
     XCOF(4) = -1.
     XCOF(5) = 1.
C
C    FIND THE ROOTS
     CALL POLRT(XCOF,COF,4,ROOTR,ROOTI,IER)
C
C    WRITE THE ANSWERS
     WRITE(6,100)
     WRITE(6,102)
 102 FORMAT(11X,'X(REAL)        X(IMAG)')
     WRITE(6,100)
     WRITE(6,101)(I,ROOTR(I),ROOTI(I),I=1,4)
 101 FORMAT(1X,I5,5X,F10.5,5X,F10.5)
 100 FORMAT(1X,35('-'))
     WRITE(6,100)
     STOP
     END
```

23

The output of this program now follows.

	X(REAL)	X(IMAG)
1	3.00000	0.
2	-4.00000	0.
3	1.00000	-3.00000
4	1.00000	3.00000

Special Methods for Finding Complex Roots

A few special methods for finding complex roots are available. These nearly always involve a procedure for extracting a quadratic factor

$$x^2 + px + q$$

from the original polynomial. One common method of this type is Lin's method. It is based on the fact that a polynomial of the form

$$x^n + a_{n-1}x^{n-1} + \ldots + a_1 x + a_0 = 0$$

can be written as:

$$0 = (x^2 + px + q)(x^{n-2} + b_{n-1}x^{n-3} + \ldots + b_3 x + b_2) + b_1 x + b_0$$

In this expression, $b_1 x + b_0$ is a linear remainder term which we desire to be zero. A zero remainder would mean that the original polynomial is exactly divisible by the quadratic factor. If it is assumed that $b_1 = b_0 = 0$ and if like coefficients are compared between the two forms of the polynomial equation, the result is:

$$b_{n-1} = a_{n-1} - p$$

$$b_{n-2} = a_{n-2} - pb_{n-1} - q$$

$$\vdots$$

$$b_{n-j} = a_{n-j} - pb_{n+1-j} - qb_{n+2-j}$$

$$\vdots$$

$$p = \frac{(a_1 - qb_3)}{b_2}$$

$$q = \frac{a_0}{b_2}$$

The procedure for the iterative process that leads to a solution is illustrated in Figure 2-9. It proceeds as follows. First, initial guesses for p and q are made. These, together with the given a_i coefficients, are used

24

to calculate b_{n-1}. The value of b_{n-1} is, in turn, used to calculate b_{n-2}, and so on until all coefficients down through b_2 have been calculated. The terms b_3, b_2, a_1, and a_0 can be used in the last two equations to get improved values of p and q—say p* and q*. If the change in p and q values is sufficiently small, the process is terminated. If the change is not small enough, the new values replace p and q and the process is repeated. As it turns out, this procedure amounts to the solution of two equations in two unknowns by direct iteration. This topic is covered in more depth in a later section of this chapter.

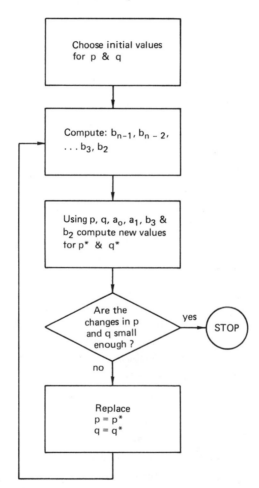

Figure 2-9   Logic flow diagram for Lin's method.

Another method based on the quadratic factor $x^2 + px + q$ uses Newton's method for the two equations and two unknowns. It is called Bairstow's method.

25

2.4   Solution of Linear Simultaneous Equations

The solution of linear simultaneous equations is, perhaps, one of the most common algebraic problems encountered in engineering calculations. The general formulation for this problem is:

$$a_{11}x_1 + a_{12}x_2 + \cdot \ \cdot \ \cdot \ \cdot + a_{1n}x_n = c_1$$

$$a_{21}x_1 + a_{22}x_2 + \cdot \ \cdot \ \cdot \ \cdot + a_{2n}x_n = c_2$$

$$\cdot \ \cdot \ \cdot \ \cdot \ \cdot \ \cdot \ \cdot \ \cdot \ \cdot \ \cdot \ \cdot \ \cdot \ \cdot \ \cdot \ \cdot \ \cdot \ \cdot \ \cdot$$

$$a_{n1}x_1 + a_{n2}x_2 + \cdot \ \cdot \ \cdot \ \cdot + a_{nn}x_n = c_n$$

In this formulation, the n-equations must be linearly independent in order for a unique solution to exist. The necessary and sufficient condition for this to occur is that the determinant of the coefficient matrix must not be zero. Solution algorithms for this type of problem may be either direct or indirect. The fact that direct methods can, for this problem situation, be systematized makes them quite popular. In this section the direct methods are considered first and the indirect (iterative) methods are considered last.

Gaussian Elimination Method

The most frequently used direct methods are based on a procedure for reducing the equation system into a "triangular" form so that one of the equations contains only one of the unknowns, and each of the next equations contains only one additional new unknown. In hand-calculation techniques, traigularization is achieved by addition and subtraction of various equations after they have been multiplied by appropriate constant factors. Although this procedure may be somewhat haphazard when implemented by hand, a systematic scheme for computer implementation can be established. The method of Gaussian elimination is one such method. This procedure starts by normalizing the first equation by dividing each of its coefficients by a_{11}. Next, this first equation is multiplied by the leading coefficient $a_{i,1}$ of each of the other equations and is subtracted from each successive equation. The result will be the elimination of the first variable from all equations except the first. Next, using the last n-1 equations and the same procedure, the second variable is eliminated from the last n-2 equations. The procedure is repeated until after n-stages the triangular form is complete. Mathematically, this procedure can be stated as follows:

At the kth stage in the elimination process, the new, normalized coefficients for the kth equation are:

26

$$b_{k,j} = \frac{a_{k,i}}{a_{k,k}},$$

and the new coefficients in the equations that follow will be:

$$b_{i,j} = a_{i,j} - a_{i,k}b_{k,j} \qquad i > k.$$

In performing this process, it should be kept in mind that the $a_{i,j}$ coefficients of the lower equations change during each stage of the process. Thus the $b_{i,j}$ coefficients for one stage become the $a_{i,j}$ coefficients that are used for the next stage. This procedue is best illustrated by a simple illustration.

Suppose that it is desired to solve the following equations by Gaussian elimination:

$$x_1 + x_2 + x_3 - x_4 = 2$$

$$x_1 - x_2 - x_3 + x_4 = 0$$

$$2x_1 + x_2 - x_3 + 2x_4 = 9$$

$$3x_1 + x_2 + 2x_3 - x_4 = 7$$

For ease of manipulation, the rows in this system will be identified by letter, and only the coefficient array will be written. The original array to be manipulated is thus:

Row	Array				
A_1	1	1	1	-1	2
A_2	1	-1	-1	1	0
A_3	2	1	-1	2	9
A_4	3	1	2	-1	7

After the elimination of x_1 terms, the array will be:

Row	Array				
$B_1 = A_1/1$	1	1	1	-1	2
$B_2 = A_2 - B_1$	0	-2	-2	2	-2
$B_3 = A_3 - 2B_1$	0	-1	-3	4	5
$B_4 = A_4 - 3B_1$	0	-2	-1	2	1

After the elimination of x_2 terms, the array will be:

Row	Array				
B_1	1	1	1	-1	2
$C_2 = B_2/(-2)$	0	1	1	-1	1
$C_3 = B_3 + C_2$	0	0	-2	3	6
$C_4 = B_4 + 2C_2$	0	0	1	0	3

27

After the elimination of x_3 terms, the array will be:

Row	Array				
B_1	1	1	1	-1	2
C_2	0	1	1	-1	1
$D_3 = C_3/(-2)$	0	0	1	-3/2	-3
$D_4 = C_4 - D_3$	0	0	0	3/2	6

After reducing the coefficients on the last row, the array will be:

Row	Array				
B_1	1	1	1	-1	2
C_2	0	1	1	-1	1
D_3	0	0	1	-3/2	-3
$E_4 = D_4/(3/2)$	0	0	0	1	4

The array can now be written in equation form as:

$$x_1 + x_2 + x_3 - x_4 = 2$$
$$x_2 + x_3 - x_4 = 1$$
$$x_3 - (3/2)x_4 = -3$$
$$x_4 = 4$$

By applying the process of back substitution, the results:

$$x_1 = 1,$$
$$x_2 = 2,$$
$$x_3 = 3, \quad \text{and}$$
$$x_4 = 4$$

can be found.

This example clearly illustrates the fact that it is desirable to reduce all off-diagonal elements to zero. This procedure is called "diagonalization" and is an improvement over triangularization.

Gauss-Jordan Elimination

The Gauss-Jordan elimination method provides a systematic means for the diagonalization of a system of linear simultaneous equations. The only mathematical difference between this direct method and the previous direct method is that $i \neq k$ is substituted for $i > k$. The kth row is called the pivot row. In the previous method, only the equations below the pivot row were manipulated, whereas in the Gauss-Jordan method, the manipulation takes place both above and below the pivot row. To illustrate this procedure, the previous example will be worked using the Gauss-Jordan procedure. The original array to be manipulated is:

Row		Array			
A_1	1	1	1	-1	2
A_2	1	-1	-1	1	0
A_3	2	1	-1	2	9
A_4	3	1	2	-1	7

After the elimination of the x_1 terms, the array will be:

Row		Array			
$B_1 = A_1/1$	1	1	1	-1	2
$B_2 = A_2 - B_1$	0	-2	-2	2	-2
$B_3 = A_3 - 2B_1$	0	-1	-3	4	5
$B_4 = A_4 - 3B_1$	0	-2	-1	2	1

Up to this point, the procedure is identical to that for the Gaussian elimination technique. After the elimination of the x_2 terms, the array will be:

Row		Array			
$C_1 = B_1 - C_2$	1	0	0	0	1
$C_2 = B_2/(-2)$	0	1	1	-1	1
$C_3 = B_3 + C_2$	0	0	-2	3	6
$C_4 = B_4 + 2C_2$	0	0	1	0	3

This new array is, of course, different from that found for the third stage of the Gaussian elimination technique. After the elimination of the x_3 terms, the array will be:

Row		Array			
$D_1 = C_1 - (0)D_3$	1	0	0	0	1
$D_2 = C_2 - D_3$	0	1	0	1/2	4
$D_3 = C_3/(-2)$	0	0	1	-3/2	-3
$D_4 = C_3 - D_3$	0	0	0	3/2	6

Finally, after the elimination of the x_4 terms in all rows except the last, the array will be:

Row		Array			
$E_1 = D_1 + (0)E_4$	1	0	0	0	1
$E_2 = D_2 - (1/2)E_4$	0	1	0	0	2
$E_3 = D_3 + (3/2)E_4$	0	0	1	0	3
$E_4 = D_4 / (3/2)$	0	0	0	1	4

Clearly, for this method the answers are easier to extract. The disadvantage to this method is its need for additional calculations.

A potential pitfall in the two foregoing methods can occur if any pivot element is zero. When this happens the normalization of the pivot row cannot be accomplished. Since it is possible to change the order of the equations in the system, this device can provide a way to circumvent the zero pivot element problem. Indeed, it can be shown that the greatest overall computational accuracy is achieved when the pivot element has the greatest magnitude. Thus the row with a zero or small pivot element should be exchanged with the row below it that has the largest element in the same column.

A number of other direct methods for triangular decomposition are available in the literature; these include the methods of Crout, Doolittle, and Choleski. Each method has certain advantages over Gaussian elimination under specific computational circumstances.

Although direct methods are normally quite efficient at providing a solution, they tend to become less efficient than indirect methods when they are applied to sparse matrices. For this reason several iterative methods for simultaneous linear equations are now presented.

Iterative Methods for Simultaneous Linear Equations

Iterative schemes for simultaneous linear equations are based on a formulation of the equations in which each of the n-variables stands alone on the left side of one of the n-equations. This form will be:

$$x_1 = b_{1,n}x_n + b_{1,n-1}x_{n-1} + \cdots + b_{1,2}x_2 + b_{1,0}$$

$$x_2 = b_{2,n}x_n + b_{2,n-1}x_{n-1} + \cdots + b_{2,1}x_1 + b_{2,0}$$

$$\vdots$$

$$x_n = b_{n,n-1}x_{n-1} + \cdots + b_{n,2}x_2 + b_{n,1}x_1 + b_{n,0}$$

The iterative techniques utilizing this formulation are those described in the chapter on partial differential equations. They are the Jacobi method, the Gauss-Seidel method, and the method of successive over-relaxation. These methods are based on the successive improvement of initial guesses for the solution.

In the Jacobi method, the initial guesses are used to generate new values for x_1 through x_n using the previous equations. If each of these new values is sufficiently close to the initial values, the process is terminated. If not, the new values replace the previous values, and the process is repeated until convergence is obtained or until it is clear that the process will never converge. In this method the replacement of solution values for all variables is done at the same time (simultaneous displacement).

In the Gauss-Seidel method, the improvement found for x_1 is used immediately in the calculation for x_2. The new values of x_1 and x_2 are then used to calculate x_3, and so on. This simple variation in the iterative process can give considerable improvement in the rate of convergence.

In the method of over-relaxation the new values computed for each variable will be:

$$x_i^{(n+1)} = x_i^{(n)} + \omega(\bar{x}_i^{(n+1)} - x_i^{(n)})$$

where $\bar{x}_i^{(n+1)}$ is the Gauss-Seidel improved value and ω is the relaxation factor that satisfies $1 \le \omega \le 2$. If $\omega = 1$, this technique reduces to the Gauss-Seidel method. The selection of the value ω will influence the rate of convergence.

2.5 Solution of Nonlinear Simultaneous Algebraic Equations

Unlike linear systems, nonlinear systems cannot be solved by direct methods because none exist. Thus the solution of nonlinear systems is always accomplished by iterative methods.

The most general formulation for nonlinear algebraic systems can be stated in the following form. Given n-functions f_i in terms of n-unknown variables x_i:

$$f_1 (x_1, x_2, \cdots, x_n) = 0$$
$$f_2 (x_1, x_2, \cdots, x_n) = 0$$
$$\vdots$$
$$f_n (x_1, x_2, \cdots, x_n) = 0$$

find the solutions. In this section we deal with common solution methods and their limitations.

Direct Iteration

The method of direct iteration for the solution of nonlinear algebraic equations is actually an extension of the method of direct iteration for

31

single equations. It is based on the assumption that the equation system can be manipulated into the form:

$$x_1{}^* = g_1(x_1, x_2, x_3, \cdots, x_n)$$

$$x_2{}^* = g_2(x_1^*, x_2, x_3, \cdots, x_n)$$

$$\vdots$$

$$x_n{}^* = g_n(x_1^*, x_2^*, x_3^*, \cdots, x_n)$$

The procedure for solution extraction is illustrated in Figure 2-10 and proceeds as follows. Using initial x_i guesses and any new values x_{i+1}^*,

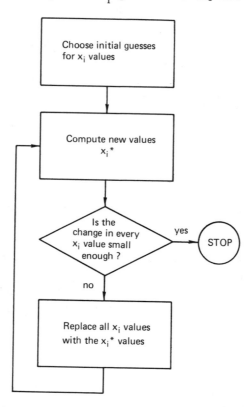

Figure 2-10   Logic flow diagram for the direct iteration method for the solution of nonlinear algebraic systems.

new values $x_1^*, x_2^*, \cdots, x_n^*$ are computed from the g_i equations. The x_i^* values are compared with the previous x_i values to see if the change is sufficiently small. If the change in every variable is small enough, the process is terminated. If the change in any variable value is too large, the process is repeated using the x_i^* values as new starting values.

32

Although this solution method is straightforward, it is not without its pitfalls. For example, if the initial guesses are not sufficiently close to the true solution, the process will fail to converge. The space domain within which an initial guess will converge to a solution is called the "domain of convergence." Any initial guesses outside of this domain will not lead to a solution. It is an unfortunate reality for nonlinear algebraic systems that, as he number of equations (and thus the number of unknowns) goes up, the size of the domain of convergence gets smaller. Indeed, for extremely large systems, the initial guesses must be very close to the actual solution values in order to achieve convergence. Although there do exist some methods for improving the convergence difficulties in problems of this type, there is no substitute for good engineering judgment in the selection process of initial values to start the iteration process.

Newton's Iteration Method

Newton's iteration method is, by far, the most commonly used method for the solution of systems of nonlinear algebraic equations. Its popularity is due to the fact that it has better convergence properties than does the method of direct iteration. The basis for Newton's iteration method is a Taylor's expansion of each of the n-equations:

$$f_1(x_1+\Delta x_1, \cdots, x_n+\Delta x_n) = f_1(x_1, \cdots, x_n) + \Delta x_1 \frac{\partial f_1}{\partial x_1}$$

$$+ \cdots + \Delta x_n \frac{\partial f_1}{\partial x_n} + \text{higher order terms}$$

$$\vdots$$

$$f_n(x_1+\Delta x_1, \cdots, x_n+\Delta x_n) = f_n(x_1, \cdots, x_n) + \Delta x_1 \frac{\partial f_n}{\partial x_1}$$

$$+ \cdots + \Delta x_n \frac{\partial f_n}{\partial x_n} + \text{higher order terms}.$$

If the Δx_i changes in the variable values bring the functions f_j close to a root, it will be assumed that the left hand sides of these equations are zero. Thus the problem reduces to that of finding the changes Δx_i that achieve the goal. If the higher order terms are dropped, the problem becomes one of finding the roots of the linear system:

$$\begin{bmatrix} \frac{\partial f_1}{\partial x_1} & \frac{\partial f_1}{\partial f_2} & \cdots & \frac{\partial f_1}{\partial x_n} \\ \cdot & & & \\ \cdot & & & \\ \cdot & & & \\ \frac{\partial f_n}{\partial x_1} & & \cdots & \frac{\partial f_n}{\partial x_n} \end{bmatrix} \begin{bmatrix} \Delta x_1 \\ \Delta x_2 \\ \cdot \\ \cdot \\ \cdot \\ \Delta x_n \end{bmatrix} = \begin{bmatrix} -f_1 \\ -f_2 \\ \cdot \\ \cdot \\ \cdot \\ -f_n \end{bmatrix}$$

33

In this system the partial derivative matrix and the vector on the right-hand side can each be evaluated at any approximate set of solution values. Once the Δx_i values are known, they may be applied as corrections to the initial approximations:

$$x_1 = x_1 + \Delta x_1$$
$$\vdots$$
$$x_n = x_n + \Delta x_n$$

If all correction factors are sufficiently small, the process is terminated. If not, the new values are used as root approximations, and the process is repeated until a solution is found or until it is obvious that no solution can be achieved. The logic flow diagram for this procedure is shown in Figure 2-11. Care should be exercised in the test for convergence. If

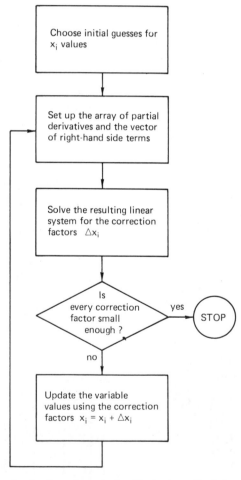

Figure 2-11 Logic flow diagram for Newton's method for the solution of nonlinear algebraic systems.

34

the individual root values x_i are of greatly different magnitudes, a check requirement such as:

$$\Delta x_i \leq 0.0001 \qquad i = 1, \ldots, n$$

may place unfair limitations on the larger x_i values. When this situation occurs it is better to test a normalized correction factor:

$$\frac{\Delta x_i}{x_i} \leq 0.0001 \qquad i = 1, \ldots, n.$$

Although the Newton iteration method is the most preferable of the iterative methods, it still can have convergence problems. The size of the domain of convergence is roughly inversely related to the degree and number of equations.

Example 2-3

Suppose that it is desired to solve the following system of four equations in four unknowns:

$$x_1 + 2x_2 + x_3 + 4x_4 = 20.700$$

$$x_1^2 + 2x_1x_2 + x_4^3 = 15.880$$

$$x_1^3 + x_3^2 + x_4 = 21.218$$

$$3x_2 + x_3x_4 = 7.900$$

The method to be used will be that of Newton iteration. In order to implement this method, it is necessary to find the partial derivatives of each of the equations in the null form. This matrix of partial derivatives is then used to compute correction factors for initial guesses for the four x_i values. The calculation of correction factors requires the solution of four linear equations in four unknowns. The following computer program implements the solution of this example problem. This program utilizes subroutine GELG from the IBM Scientific Subroutine Package to solve the linear system of equations by the method of Gauss elimination. The initial guesses used to start the iterative process are $x_1 = x_2 = x_3 = x_4 = 1.0$.

```
C       :************************************************
C       *      THIS PROGRAM SOLVES A SYSTEM OF          *
C       *      4 NONLINEAR ALGEBRAIC EQUATIONS FOR      *
C       *      4 UNKNOWNS BY THE METHOD OF NEWTON       *
C       *      ITERATION.  THIS PROGRAM MAKES USE OF    *
C       *      SUBROUTINE GELG FROM THE IBM SCIENTIFIC  *
C       *      SUBROUTINE PACKAGE.                      *
C       *                    T. E. SHOUP 10/25/77       *
C       ************************************************
        DIMENSION X(4),V(4,4),G(4)
C
C       LOAD THE INITIAL VALUES
        DO 1 I=1,4
     1  X(I) = 1.
```

35

```
          WRITE(6,101)
  101     FORMAT(1X,58('-'))
          WRITE(6,102)
  102     FORMAT(1X,'ITERATION',8X,'X(1)',8X,'X(2)',8X,'X(3)',8X,'X(4)')
          WRITE(6,101)
          I = 0
          WRITE(6,100)I,(X(II),II=1,4)
          DO 2 I=1,50
C         LOAD THE PARTIAL DERIVATIVE ARRAY
          V(1,1) = 1.0
          V(1,2) = 2.0
          V(1,3) = 1.0
          V(1,4) = 4.0
          V(2,1) = 2.*X(1) + 2.*X(2)
          V(2,2) = 2.*X(1)
          V(2,3) = 0.
          V(2,4) = 3.*X(4)**2
          V(3,1) = 3.*X(1)**2
          V(3,2) = 0.0
          V(3,3) = 2.*X(3)
          V(3,4) = 1.0
          V(4,1) = 0.
          V(4,2) = 3.
          V(4,3) = X(4)
          V(4,4) = X(3)
C
C         LOAD THE EQUATION VECTOR
          G(1) = -1.*X(1)-2.*X(2)-X(3)-4.*X(4)+20.70
          G(2) = -1.*X(1)**2-2.*X(1)*X(2)-X(4)**3+15.88
          G(3) = 21.218-X(1)**3-X(3)**2-X(4)
          G(4) = -3.*X(2) -X(3)*X(4)+21.1
          DO 3 J=1,4
  3       IF(ABS(G(J)).GT.0.00001) GO TO 4
          GO TO 6
  4       CALL GELG(G,V,4,1,0.000001,IER)
C
C         UPDATE THE X(I) VALUES
          DO 5 K=1,4
  5       X(K) = X(K) + G(K)
  2       WRITE(6,100)I,(X(II),II=1,4)
  100     FORMAT(5X,I5,4(2X,F10.5))
  6       WRITE(6,101)
          STOP
          END
```

In this program the iterative process is terminated if the residual equations get smaller than 0.00001 or if the number of iterations exceeds 50. The output of this program is:

ITERATION	X(1)	X(2)	X(3)	X(4)
0	1.00000	1.00000	1.00000	1.00000
1	2.75037	4.67630	7.89579	0.17531
2	1.34485	5.29712	5.94935	0.70289
3	1.47750	3.84372	4.34185	1.79830
4	1.54266	6.24338	4.12036	0.63755
5	1.23637	5.72736	4.34362	0.91632
6	1.20239	5.59862	4.29949	1.00022
7	1.20000	5.60000	4.30000	1.00000
8	1.20000	5.60000	4.30000	1.00000

Parameter Perturbation Procedure

The parameter perturbation procedure is an algorithm for enabling an iterative solution method to achieve a root to a set of simultaneous non-linear equations. It is a procedure that operates independently of the need for a "good" initial approximation. The procedure is implemented as follows. First, given a set of equations:

$$f_j \ (x_i) = 0 \qquad\qquad i = 1, \ . \ . \ . \ , \ n$$

$$j = 1, \ . \ . \ . \ , \ n$$

Consider another set of equations:

$$g_j \ (x_i) = 0 \qquad\qquad i = 1, \ . \ . \ . \ , \ n$$

$$j = 1, \ . \ . \ . \ , \ n$$

for which a solution set is known. The $g_j = 0$ equations are "deformed" into the equations $f_j = 0$ by means of a finite number "N" of successive small increments in the parameters:

$$g_j^{(k)}(x_i) = g_j^{(k-1)}(x_i) + [f_j \ (x_i) - g_j^{(k-1)} \ (x_i)] \ \frac{k}{N}$$

$$k = 1, \ . \ . \ . \ , \ N$$

The initial solution set $x_i^{(0)}$ to the initial set of equations $g_j^{(0)} \ (x_i)$ can be used as an initial guess for the iterative solution of the $g_j^{(1)} \ (x_i)$ set. Since this set is only slightly different from the previous set, convergence is likely. As the procedure progresses, the $x_i^{(k-1)}$ solution becomes the initial guess to obtain the $x_i^{(k)}$ solution. In the end, when k = N the system is equivalent to the original set of equations. Since it may require N = 10 or N = 100 steps to deform the known problem into the desired problem, the procedure can require considerable computer time to implement. Fortunately, if the step sizes are small, convergence at any single step can often be accomplished with only a few iterations.

The parameter perturbation procedure has been shown to be especially useful for the solution of equation systems encountered in the synthesis of mechanical linkages. In such cases any arbitrary mechanism of the type required can serve to generate the starting system $g_j^{(0)}(x_i) = 0$.

2.6   Considerations in the Selection of a Method for the Solution of   Algebraic Systems

Nearly every computer software package in existence includes some methods for handling algebraic problems. Thus to list even a representative sample would be futile. However, it is intended here to make some general statements concerning the utility of a particular algorithm in a given situation. These statements are categorized by the type of problem to which they apply.

Transcendental Equations
In solving transcendental equations, it should be kept in mind that the Newton's method, the direct iteration method, and the secant method are often faster than other methods. Unfortunately, these methods do not always converge. On the other hand, the slower methods such as the binary search method and the method of false position will guarantee a solution for any continuous function once a function sign change has been found.

Polynomial Equations
Of all of the iterative methods available for the solution of a polynomial of order "n," some are better for a particular problem situation than others. Ralston and Wilf [7] recommend the following methods for highest efficiency:

order	method
$3 \leq n \leq 5$	Newton
$6 \leq n \leq 84$	Secant
$85 \leq n$	Other special methods

Simultaneous Linear Equations
For general systems of simultaneous linear equations, the elimination methods with pivoting are best to use. Iterative methods are efficient only for sparse matrix problems.

Simultaneous Nonlinear Equation Systems
The method of Newton iteration is the better method for solving simultaneous nonlinear equation systems. In the event that the derivatives cannot be found, it is possible to use a secant approximation for the partial derivatives.

Problems

2.1 Find solutions to the following transcendental equations:

$$a) \quad 0.25X - \sin(X) = 0$$

$$b) \quad \sin(Y) - 1/Y = 0$$

38

c) $\ln(Z) - \sin(Z) = 0$

d) $4X - \cosh(X) = 0$

e) $Y^2 - \operatorname{sech}(Y) = 0$

2.2   Find at least one real root to the following polynomial equations:

a) $x^4 + 7x^3 + 3x^2 + 4x + 1 = 0$

b) $7x^4 + 5x^3 + 2x^2 + 4x + 1 = 0$

c) $x^4 + 5x^3 + 5x^2 - 5x + 6 = 0$

d) $x^5 + x^4 + 2x^2 - x - 2 = 0$

2.3   Find all roots to the following polynomial equations:

a) $x^4 + 8x^3 + 2x^2 + 3x + 1 = 0$

b) $x^4 + 2x^3 + 7x^2 + 2x + 7 = 0$

c) $x^5 + 3x^4 + 6x^2 + 3 + 1 = 0$

d) $x^7 + 3x^3 + 2x + 1 = 0$

e) $x^6 + 2x^4 + 3x^2 + 5 = 0$

f) $x^4 + 3x^3 + 2x^2 + 5x = 0$

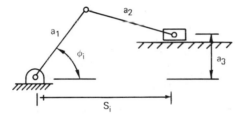

2.4   The slider crank mechanism shown above is described by the equation:

$$K_1 s_i \cos(\phi_i) + K_2 \sin(\phi_i) - K_3 = s_i^2 \qquad \text{for } i=1,2,3$$

where:

$$a_1 = \frac{K_1}{2}, \quad a_3 = \frac{K_2}{2a_1} \quad \text{and} \quad a_2 = \sqrt{a_1^2 + a_3^2 - K_3}$$

It is desired to use this device to satisfy the following conditions:

i	s_i	ϕ_i
1	1.0	20°
2	1.2	45°
3	2.0	60°

Design a device that will satisfy all three of these positions by writing the describing equation three times and solving for the K_i values. What are the sizes of a_1, a_2 and a_3 for this solution?

2.5 Find a solution to the following system of nonlinear algebraic equations:

$$x_1 \quad + x_2 \quad + x_3 \quad + x_4 = 5.0$$

$$x_1^2 \quad + x_2 \quad + x_3^2 \quad + x_4 = 12.0$$

$$x_1 x_2 + x_2 x_3 + x_4 \qquad = 5.0$$

$$x_1 x_3 + x_2 x_4 + x_4^2 \qquad = 9.0$$

2.6 The fourth order polynomial:

$$f(x) = x^4 - 12x^3 + 46x^2 - 60x + 25,$$

has a double root at $x = 1$. Try several of the techniques presented in Section 2.2 to find this root. Since the slope of the polynomial curve is zero at a double root, a convenient way to find double roots when they are present is to first solve:

$$\frac{df}{dx} = 0$$

and then see if any of these roots are also roots to the original equation:

$$f(x) = 0.$$

Try this method for this polynomial.

2.7 The root locus method of illustrating the relative stability of a control system is a way to determine how the effect of increasing system gain will influence system behavior. Since positive real parts to the roots of the system characteristic equation will give rise to transient behavior that grows exponentially, this situation is to be avoided at all costs. For the control system:

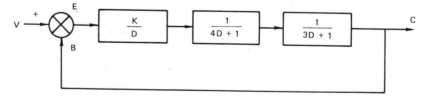

the characteristic equation will be:

$$12D^3 + 7D^2 + D + K = 0$$

Find and plot the roots D_i in the complex plane for increasing values of K. At what value of gain K does the system go unstable?

2.8 The system of equations:

$$5x_1 + 3x_2 + x_3 + x_4 = 16$$

$$x_1 x_2 + x_2 x_3 + x_3 x_4 = 17$$

$$x_1^2 + x_2^2 + x_3^2 - x_4^2 = 9$$

$$x_1 x_3 + x_2 x_4 + x_1{}^3 \quad = 8$$

has a solution:

$$x_1 = x_2 = 1.0$$

$$x_3 = 4.0$$

$$x_4 = 3.0$$

Can you find this root by Newton's iteration method? What happens to the solution? Can you suggest an approach that will overcome this difficulty?

2.9 For the system of equations shown below:

$$x_1 + x_2 + x_3 + x_4 \qquad = 31$$

$$x_1 x_2 + x_2 x_3 + x_4 x_5 \qquad = 58$$

$$x_1{}^2 + x_3 x_4 - x_2{}^2 + x_1 x_5 \qquad = 79$$

$$x_1 - x_2 x_4 + x_3{}^2 + x_5{}^3 \qquad = 17$$

$$x_1 x_3 - x_2{}^3 x_5 - x_5 x_2 + x_3{}^2 x_4 = 234$$

Find a solution starting from $x_1 = x_2 = x_3 = x_4 = x_5 = 1.0$ using parameter perturbation by solving the sequence of systems:

$$x_1 + x_2 + x_3 + x_4 \qquad = 5 + \quad 26 \ N/10$$

$$x_1 x_2 + x_2 x_3 + x_4 x_5 \qquad = 1 + \quad 57 \ N/10$$

$$x_1{}^2 + x_3 x_4 - x_2{}^2 + x_1 x_5 \qquad = 2 + \quad 77 \ N/10$$

$$x_1 - x_2 x_4 + x_3{}^2 + x_5^3 \qquad = 2 + \quad 15 \ N/10$$

$$x_1 x_3 - x_2{}^3 x_5 - x_5 x_2 + x_3{}^2 x_4 \quad = 0 + 234 \ N/10$$

$$N = 1, 2, \ldots, 10$$

2.10 In the chemical reaction:

$$CO + \tfrac{1}{2} O_2 \overset{\rightarrow}{\leftarrow} CO_2$$

the percentage of dissociation (X) of 1 mole of CO_2 will depend on the equation:

$$(P/K^2 - 1)X^3 + 3X - 2 = 0$$

where P is the pressure on the CO_2 in atmospheres and K is the equilibrium constant[†] that depends on temperature. Find X for K = 1.648 (at 2800 °K) and P = 1 atm.

[†]Lewis, B., and von Elbe, G., "Heat Capacities and Dissociation Equilibria of Gases," J. Am. Chem. Soc. 57 (1935):612

References

1. Freudenstein, F., and Roth, B., "Numerical Solution of Systems of Nonlinear Equations." J. of the Assoc. for Compu. Mach., 10, 4 (1963): 550-56.

2. Grove, Wendell E. Brief Numerical Methods. Englewood Cliffs, N.J.: Prentice-Hall, Inc., 1966.

3. La Fara, Robert L. Computer Methods for Science and Engineering. Rochelle Park, N.J.: Hayden Book Co., 1973.

4. McCalla, Thomas Richard, Introduction to Numerical Methods and FORTRAN Programming. New York: John Wiley & Sons, 1967.

5. Pall, Gabriel A. Introduction to Scientific Computing. New York: Appleton-Century-Crofts, Educational Division, Meredith Corp., 1971.

6. Ralston, Anthony, A First Course in Numerical Analysis. New York: McGraw-Hill Book Co., 1965.

7. Ralston, Anthony and Wilf, H.S. Mathematical Methods for Digital Computers. New York: John Wiley & Sons, 1967.

8. Salvadori, Mario G., and Baron, Melvin L. Numerical Methods in Engineering. Englewood Cliffs, N.J.: Prentice-Hall, Inc., 1961.

9. Turner, L. R. "Solution of Nonlinear Systems." Annals New York Academy of Sciences, 86 (1960): 817-27.

10. Williams, P. W. Numerical Computation. New York: Barnes & Noble Import Division, Harper & Row, 1972.

3 COMPUTER SOLUTION OF EIGENVALUE PROBLEMS

In the industrial environment the computer allows the instant check-out of the transient characteristics of large rotating machines. [Photo courtesy of Westinghouse Electric Corp.]

3.1　Introduction

The analysis of some engineering problems leads to sets of algebraic equations that can have a unique solution only when the value of a parameter within the equations is known. This special parameter is called a characteristic value or eigenvalue. The solution associated with each eigenvalue is called its eigenvector. Eigenvalue problems occur in a variety of situations. In the manipulation of stress tensors, the eigenvalues identify the principal normal stresses, and the eigenvectors identify the orientations associated with these values. In the dynamic analysis of systems, the eigenvalues identify the natural frequencies of vibration, and the eigenvectors characterize the shapes of these vibration modes. In structural analysis, eigenvalues can be used to determine the critical buckling loads.

Selection of the best technique for finding the eigenvalues and eigenvectors for a given engineering problem will depend on several factors such as the nature of the equations, the number of eigenvalues desired, and the nature of these eigenvalues. In general there are two categories of solution algorithms for eigenvalue problems. The iterative methods are quite easy to use and are well suited for finding the smallest and largest eigenvalues. The transformation methods are a bit more difficult to apply but have the advantage that they identify all eigenvalues and eigenvectors.

It is the purpose of this chapter to discuss the common techniques available for solution of eigenvalue problems. Before presenting these techniques, it will be useful to review some fundamentals of matrix and vector theory upon which the eigenvalue methods are based.

3.2　Fundamentals of Eigenvalue Problems

The general formulation of an eigenvalue problem is:

$$AX = \lambda X$$

where A is an n x n matrix. In this expression it is desired to find the n scalar values of λ and the eigenvectors X associated with each of these eigenvalues.

Although it is assumed that the reader has a basic knowledge of matrix calculus and of eigenvalue theory, a few fundamental principles are worthy of reemphasis before considering the solution methods.

Fundamentals of Matrices
1.　A matrix A is symmetric if
$$a_{ij} = a_{ji} (i,j = 1,2,..,n)$$

44

This gives rise to symmetry about the diagonal

$$a_{kk} \quad (k = 1, 2, .., n).$$

An example of a symmetric matrix would be:

2. A matrix A is said to be tridiagonal if all elements are zero except the main diagonal elements and the diagonal elements immediately above and below the main diagonal. The general form of a tridiagonal matrix is:

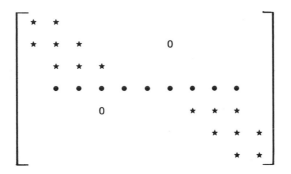

The tridiagonal form is important because some transformation methods reduce a general matrix to this special form.

3. A matrix is orthogonal if

$$A^T A = I$$

where A^T represents the transpose of the A matrix and I represents the identity matrix. Clearly, the inverse of an orthogonal matrix is equivalent to its transpose.

4. Two matrices A and B are said to be similar if a non-singular matrix P can be found such that

$$B = P^{-1} AP.$$

Fundamentals of Eigenvalues

1. The n eigenvalues of any real symmetric n x n matrix will all be real. This fact is particularly useful because many engineering matrices are symmetrical.

2. If the eigenvalues of a matrix are distinct, then the eigenvectors will be orthogonal. A set of n linearly independent eigenvectors will form a base for the space being considered. This means that for the linearly independent set of eigenvectors

45

$$X^i \quad i = 1, \ldots, n$$

any arbitrary vector Y in the space can be expressed in terms of the eigenvectors. Thus

$$Y = \sum_{i=1}^{n} a_i X^i.$$

3. If two matrices are similar, they have the same eigenvalues. Thus the similarity of A and B means:

$$B = P^{-1}AP.$$

Since

$$AX = \lambda X$$

then

$$P^{-1} AX = \lambda P^{-1}X.$$

If one lets X = PY, then:

$$P^{-1} APY = \lambda Y$$

and

$$BY = \lambda Y.$$

Thus not only do the two matrices have the same eigenvalues but their eigenvectors are related by X = PY.

4. Any scalar multiple of an eigenvector of a matrix is also an eigenvector of that matrix. It is customary to normalize all eigenvectors either by dividing each element of the eigenvector by the largest element or by dividing each element by the sum of the squares of the other elements.

3.3 Iterative Methods of Solution

Perhaps the most obvious technique for the solution of eigenvalues can be found by expressing the eigenvalue problem as:

$$(A - \lambda I)X = 0.$$

This system will have a nonzero solution only if the determinant of $(A - \lambda I)$ is zero. This determinant will give a polynomial in λ of degree n, and the roots of this polynomial will be the eigenvalues. Thus any of the methods presented in Chapter 2 may be applied to find these roots. Unfortunately, eigenvalue problems often have multiple roots. Because the iterative methods of Chapter 2 do not work well for this situation, it is best to utilize other iterative procedures to extract the eigenvalues.

Finding the Largest Eigenvalue by Iteration

The basic iterative method for finding the largest eigenvalue of the system:

$$AX = \lambda X$$

is illustrated in Figure 3-1. The procedure begins with a trial normalized

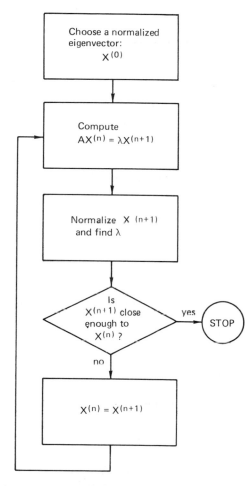

Figure 3-1   Logic flow diagram for the iterative method of eigenvalue determination.

vector $X^{(0)}$. This vector is multiplied by the A matrix on the left-hand side to obtain the product λX. Next, the product λX is reduced to a constant (the eigenvalue) and a normalized vector $X^{(1)}$. If the vector $X^{(1)}$ reproduces the vector $X^{(0)}$, the process may be terminated. If not, the new normalized vector becomes the starting vector, and the process is

47

repeated. When the process has converged, the constant multiplier will be the correct value for the largest eigenvalue, and the normalized vector will be the corresponding eigenvector. The rate of convergence for this iterative process depends on the choice of the initial vector. A choice close to the actual eigenvector gives rapid convergence. The rate of convergence is also influenced by the ratio of the sizes of the two largest eigenvalues. When this ratio is nearly unity, the convergence rate is poor.

Example 3-1

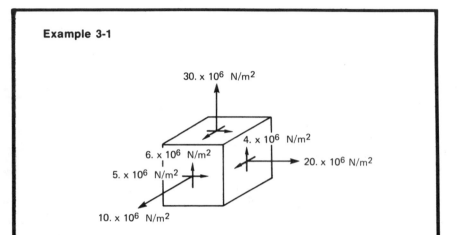

30. x 10^6 N/m²
4. x 10^6 N/m²
6. x 10^6 N/m²
20. x 10^6 N/m²
5. x 10^6 N/m²
10. x 10^6 N/m²

It is desired to investigate the triaxial state of stress indicated in the figure above. For this elemental block, the stress matrix will be:

$$\begin{bmatrix} 10. & 5. & 6. \\ 5. & 20. & 4. \\ 6. & 4. & 30. \end{bmatrix} \text{ x } 10^6 \quad \text{N/m}^2$$

If the maximum stress theory of failure is to be used for this investigation, it will be necessary to know the largest principal stress value. This stress corresponds to the largest eigenvalue of the stress matrix. Thus an iterative method will be used. The following computer program which implements this iterative procedure, iterates until the change in the eigenvalue becomes smaller than 0.01%. The program utilizes two subroutines. Subroutine GMPRD is a routine from the IBM Scientific Subroutine Package that performs the required matrix multiplication. The subroutine NORML normalizes the eigenvectors with respect to the largest element.

```
C     *******************************************************
C     *        E I G E N V A L U E     P R O G R A M        *
C     *     THIS PROGRAM FINDS THE LARGEST PRINCIPAL STRESS *
C     *     (EIGENVALUE) FOR A GIVEN TRIAXIAL STATE OF      *
C     *     STRESS.  THE METHOD USED IS THAT OF ITERATION.  *
C     *     THE ITERATION PROCESS IS STOPPED WHEN THE       *
C     *     CHANGE IN EIGENVALUE BECOMES SMALLER THAN       *
```

```
C     *       0.01 PERCENT OR THE NUMBER OF ITERATIONS         *
C     *       EXCEEDS 50.                                      *
C     *                               T. E. SHOUP   10/10/77   *
C     ********************************************************
      DIMENSION S(3,3),X(3),R(3)
      S(1,1) = 10.E06
      S(1,2) = 5.E06
      S(2,1) = S(1,2)
      S(1,3) = 6.E06
      S(3,1) = S (1,3)
      S(2,2) = 20.E06
      S(2,3) = 4.E06
      S(3,2) = S(2,3)
      S(3,3) = 30.E06
      X(1) = 1.
      X(2) = 0.0
      X(3) = 0.0
      XOLD = 0.0
      I = 0
      WRITE(6,100)
      WRITE(6,101)
      WRITE(6,102)
      WRITE(6,100)
      WRITE(6,104) I,X(1),X(2),X(3)
      DO 1 I=1,50
      CALL GMPRD (S, X, R, 3, 3, 1)
      DO 2 J=1,3
    2 X(J) = R(J)
      CALL NORML(XLAM,X)
      WRITE(6,103) I,XLAM,X(1),X(2),X(3)
      IF(ABS((XOLD-XLAM)/XLAM).LE.0.0001) GO TO 3
    1 XOLD = XLAM
    3 WRITE(6,100)
  100 FORMAT(1X,54('-'))
  101 FORMAT(2X,'ITERATION',3X,'EIGENVALUE',11X,'EIGENVECTOR')
  102 FORMAT(3X,'NUMBER',6X,'(N/M**2)',5X,'X(1)',6X,'X(2)',6X,'X(3)')
  103 FORMAT(1X,I5,7X,E12.5,3F10.5)
  104 FORMAT(1X,I5,19X,3F10.5)
      STOP
      END



      SUBROUTINE NORML(XL,X)
      DIMENSION X(3)
C     ********************************************************
C     *             S U B R O U T I N E    N O R M L         *
C     *     THIS SUBROUTINE FINDS THE LARGEST OF THREE       *
C     *     EIGENVECTOR ELEMENTS AND THEN NORMALIZES THE     *
C     *     EIGENVECTOR WITH RESPECT TO THIS LARGEST         *
C     *     ELEMENT.                                         *
C     *                               T. E. SHOUP   10/10/77 *
C     ********************************************************
C     FIND THE LARGEST ELEMENT
      XBIG = X(1)
      IF(X(2).GT.XBIG)XBIG=X(2)
      IF(X(3).GT.XBIG)XBIG=X(3)
C     NORMALIZE WITH RESPECT TO XBIG
      X(1) = X(1)/XBIG
      X(2) = X(2)/XBIG
      X(3) = X(3)/XBIG
      XL = XBIG
      RETURN
      END
```

The output of this program now follows. Note that it requires 14 iterations to achieve the desired degree of accuracy.

ITERATION NUMBER	EIGENVALUE (N/M**2)	EIGENVECTOR X(1)	X(2)	X(3)
0		1.00000	0.	0.
1	0.10000E 08	1.00000	0.50000	0.60000
2	0.26000E 08	0.61923	0.66923	1.00000
3	0.36392E 08	0.42697	0.56278	1.00000
4	0.34813E 08	0.37583	0.49954	1.00000
5	0.34253E 08	0.35781	0.46331	1.00000
6	0.34000E 08	0.34984	0.44280	1.00000
7	0.33870E 08	0.34580	0.43121	1.00000
8	0.33800E 08	0.34362	0.42466	1.00000
9	0.33760E 08	0.34240	0.42094	1.00000
10	0.33738E 08	0.34171	0.41884	1.00000
11	0.33726E 08	0.34132	0.41765	1.00000
12	0.33719E 08	0.34110	0.41697	1.00000
13	0.33714E 08	0.34098	0.41658	1.00000
14	0.33712E 08	0.34091	0.41636	1.00000

Finding the Smallest Eigenvalue by Iteration

In some engineering problems, it is more convenient to be able to find the smallest eigenvalue rather than the largest. This can be accomplished by premultiplying the original system by the inverse of the A matrix. Thus:

$$A^{-1}AX = \lambda A^{-1}X.$$

If both sides are then multiplied by $1/\lambda$, the result will be:

$$\frac{1}{\lambda}X = A^{-1}X.$$

Clearly, this is a different eigenvalue problem where $1/\lambda$ is the eigenvalue and A^{-1} is the matrix of interest. The maximum value of $1/\lambda$ will occur when λ is smallest. Thus the previous iterative scheme can be applied to this new system to identify the smallest eigenvalue.

Finding Intermediate Eigenvalues by Iteration

After one has found the largest eigenvalue, it is possible to find the next largest eigenvalue by replacing the original matrix with one that possesses only the remaining eigenvalues. This process of removing the largest known eigenvalue is known as deflation. For an original symmetrical matrix A with known largest eigenvalue λ_1 and eigenvector X_1, the principle of orthogonality of the eigenvectors can be invoked. This means:

$$X_i^T X_j = 0 \quad \text{for } i \neq j$$

$$= 1 \quad \text{for } i = j$$

If a new matrix A^* is formed by

$$A^* = A - \lambda_1 X_1 X_1^T,$$

the eigenvalues and eigenvectors of this new matrix will be found from:

$$A^* X_i = \lambda_i X_i.$$

From the previous expression for A^*, it is clear that:

$$A^* X_i = A X_i - \lambda_1 X_1 X_1^T X_i$$

In this expression, if $i = 1$ the orthogonality relationship requires that the right-hand side become:

$$A X_1 - \lambda_1 X_1,$$

but this must be zero since it is the definition of the eigenvalues of A. Thus the λ_1 eigenvalue of the A^* matrix is zero, and all other eigenvalues of A^* are the same as for A. Thus the eigenvalues of A^* are 0, λ_2, λ_3, .., λ_n having corresponding eigenvectors X_1, X_2, ... X_n. The larger eigenvalue λ_1 has thus been removed, and the traditional iteration method can be applied to A^* to find the next largest eigenvalue λ_2. Once λ_2 and X_2 are known, the process may be repeated using a new matrix A^{**} found from A^*, λ_2 and X_2. Although this process may appear to have considerable promise, it does have fundamental drawbacks. As each new step is performed, any errors in the eigenvectors will be passed on to the next eigenvector to cause increasing inaccuracy. For this reason the method is of questionable value for use in finding more than three eigenvalues removed from the highest or lowest. When more than this number of eigenvalues is desired, it is better to utilize similarity transformation methods.

3.4   Transform Methods of Eigenvalue Calculation

The objective in utilizing a similarity transformation method of a matrix is to obtain a new matrix having the same eigenvalues but a simpler form. Obviously, the best possible simplification would be to reduce the matrix to a pure diagonal form, because the eigenvalues would then be available by inspection of the diagonal elements. Unfortunately, most transformation methods do not achieve this special form. In general, we are satisfied if the matrix is reduced to a simplified form such as the tridiagonal form.

Jacobi's Method

Jacobi's method is designed to produce a diagonal form by eliminating each off-diagonal element in a systematic fashion. Unfortunately, the process requires an infinite number of steps to achieve a perfect diagonal form, because the introduction of a new zero term at one point in a matrix will often introduce a nonzero element into a previous zero position. In practice the Jacobi method can be viewed as an iterative procedure that eventually approaches the diagonal form close enough to be terminated. For a real symmetric matrix A, the computation is performed using orthogonal matrices that are real plane rotations. The computational scheme proceeds as follows. A new matrix A_1 is formed from the original A matrix by means of the transformation $A_1 = P_1 A P_1^T$. The orthogonal matrix P_1 is chosen so that it introduces a zero off-diagonal element in A_1. Next, a new matrix A_2 is formed from A_1 using a second transformation matrix P_2 selected so that a different off-diagonal element of A_2 will be zero. The process is continued in such a way that, at each step, the off-diagonal element of maximum magnitude is reduced to zero. The transformation matrix to accomplish this process at each stage is constructed as follows. If the $a_{k\ell}$ element of A_{m-1} has maximum magnitude, then P_m corresponds to

$$P_{kk} = P_{\ell\ell} = \cos \theta$$

$$P_{k\ell} = -P_{\ell k} = \sin \theta$$

$$P_{ii} = 1 \ (i \neq k,\ell); \ P_{ij} = 0 \text{ otherwise.}$$

The A_m matrix will differ from the A_{m-1} matrix only in rows and columns k and ℓ. In order to ensure that the $a_{k\ell}^{(m)}$ element is zero, the value of θ is selected so that

$$\tan 2\theta = \frac{2a_{k\ell}^{(m-1)}}{a_{kk}^{(m-1)} - a_{\ell\ell}^{(m-1)}}.$$

The value of θ is restricted to:

$$-\frac{\pi}{4} \leqq \theta \leqq \frac{\pi}{4}.$$

Thus the orthogonal matrix P_m will appear as:

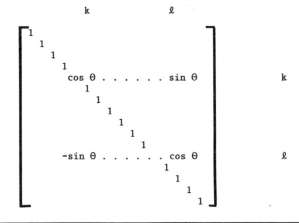

$$P_m =$$

(matrix with k and ℓ labels, containing 1's on diagonal with $\cos \theta \ldots \ldots \sin \theta$ in row k and $-\sin \theta \ldots \ldots \cos \theta$ in row ℓ)

Example 3-2

Suppose that it is desired to find all principal stress values for the state of stress presented in Example 3-1. This means that all eigenvalues must be found. The need for all eigenvalues might arise if the designer were planning to use some failure theory other than the maximum normal stress theory of failure. Since all eigenvalues are required, the transformation method of Jacobi will be used. A readily available routine that implements this method for a symmetric matrix is the subroutine EIGEN from the IBM Scientific Subroutine Package. Since the matrix is symmetrical, there are only six unique terms in the matrix. In order to save space, the subroutine EIGEN handles the 3 x 3 matrix in a compact form that requires only six storage locations. The program that solves this problem is:

```
C     ******************************************************
C     *     THIS PROGRAM FINDS ALL PRINCIPAL STRESSES      *
C     *     FROM A TRIAXIAL STRESS MATRIX.  THE            *
C     *     PROGRAM USES SUBROUTINE EIGEN FROM THE         *
C     *     IBM SCIENTIFIC SUBROUTINE PACKAGE.             *
C     *             T. E. SHOUP   11/1/77                  *
C     ******************************************************
      DIMENSION S(6),R(9)
C
C     LOAD THE STRESS MATRIX IN COMPACT FORM
      S(1) = 10.E06
      S(2) =  5.E06
      S(3) = 20.E06
      S(4) =  6.E06
      S(5) =  4.E06
      S(6) = 30.E06
C
C     FIND ALL EIGENVALUES BY THE JACOBI METHOD
      CALL EIGEN(S,R,3,0)
C
C     WRITE THE EIGENVALUES
      WRITE(6,100)
      WRITE(6,101) S(1),S(3),S(6)
  100 FORMAT(1X,'THE EIGENVALUES ARE')
  101 FORMAT(1X,E15.8)
      STOP
      END
```

The output of this program is:

```
THE EIGENVALUES ARE
 0.33709179E 08
 0.19149061E 08
 0.71417603E 07
```

Given's Method for Symmetric Matrices

Given's method is based on similarity transformations of the same type as used for Jacobi's method; however, the scheme is arranged so that once zeros are created in an element, they are retained in later transformations. For this reason the method requires only a fixed finite number of trans-formations. Thus the method is more efficient in computer time when com-pared with the Jacobi method. The only disadvantage to the method is that it produces a tridiagonal form for a symmetric matrix rather than a diagonal form. We shall see later that the tridiagonal form is quite useful and is well worth achieving.

When applied to an n x n matrix, the Given's method uses n-2 major steps. Each of these major steps may require several transformations depending on the number of zeros that must be introduced into a given column or row. In the kth step, zeros are introduced into the non-tridiagonal elements of the kth row and k^{th} column without destroying the zeros produced during the previous k-1 steps. Thus as it begins the kth step, the transformed matrix is tridiagonal as far as its first k-1 rows and columns are concerned. The pattern of transformed matrices for a symmetric 5 x 5 system would be:

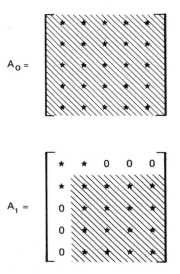

$A_0 =$ (the starting matrix)

$A_1 =$ (after the first major step consisting of three separate transformations)

54

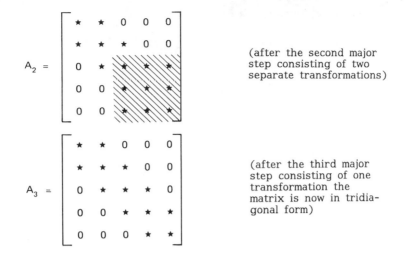

$A_2 =$ (after the second major step consisting of two separate transformations)

$A_3 =$ (after the third major step consisting of one transformation the matrix is now in tridiagonal form)

In each major step, only elements a_{ij} that are in the lower right hand shaded area are changed. Thus in the kth step, we are concerned only with the matrix of order $(n - k + 1)$ in the bottom right-hand corner of the previous matrix. Clearly, as the stages proceed, the number of transformations per stage decreases. The total number of transformations required to reach the tridiagonal form will be $(n^2 - 3n + 2)/2$.

Based on our observations of Given's method, it seems reasonable to desire a method that will reduce a whole row and column of nontridiagonal terms to zero at the same time. A method that accomplishes this has been presented by Householder.

Householder's Method for Symmetric Matrices

Because it transforms whole rows and columns of nontridiagonal elements to zero simultaneously, the method of Householder achieves a tridiagonal form in about one-half as many computations. Even though Householder's method uses a more complex transformation, it is usually faster than Given's method owing to the reduced number of transformations necessary. This efficiency is particularly obvious for large matrices. Although it utilizes Hermitian orthogonal transformation matrices rather than plane rotations, the tridiagonal form generated by Householder's method will have the same eigenvalues as the tridiagonal form generated by Given's method. The transformations involved in the n-2 major steps of this method are:

$$A_k = P_k A_{k-1} P_k^T \qquad k = 1, \ldots, n-2,$$

where

$$A_0 = A.$$

The transformation matrices will each be of the form:

$$P_k = I - \frac{u_k u_k^T}{2K_k^2}$$

where:

$$u_{i,k} = 0 \qquad\qquad i = 1,\ 2,\ldots,k$$

$$u_{i,k} = a_{k,i} \qquad\qquad i = k+2,\ldots,n$$

$$u_{k+1,k} = a_{k,k+1} \mp S_k$$

In these elements:

and

$$S_k = \left[\sum_{i=k+1}^{n} a_{k,i}^2 \right]^{1/2}$$

$$2K_k^2 = S_k^2 \mp a_{k,k+1}S_k.$$

The sign choice in these equations is taken to be the same as that for $a_{k,k+1}$. In this way the value of $u_{k+1,k}$ is maximized. It should be noted that the Given's and Householder's methods can be applied to non-symmetric matrices. The result will not be a tridiagonal form but will be a special form known as the Hessenberg form. The Hessenberg form is triangular matrix of the form:

$$\begin{bmatrix} \star & \star & 0 & 0 & 0 & 0 \\ \star & \star & \star & 0 & 0 & 0 \\ \star & \star & \star & \star & 0 & 0 \\ \star & \star & \star & \star & \star & 0 \\ \star & \star & \star & \star & \star & \star \\ \star & \star & \star & \star & \star & \star \end{bmatrix}$$

3.5 Finding the Eigenvalues of a Symmetric Tridiagonal Matrix

Once Given's or Householder's method has been applied to a symmetric matrix and a tridiagonal form has been achieved, it next becomes necessary to find the eigenvalues. In order to see the utility of the tridiagonal form, the eigenvalue problem will be written in the form:

$$\det(A - \lambda I) = 0$$

where A is a symmetric tridiagonal matrix. This system will look as follows:

$$\det \begin{vmatrix} a_1 - \lambda & b_2 & & & & \\ & & & & 0 & \\ b_2 & a_2 - \lambda & & & & \\ & & \ddots & & & \\ & & & & & b_n \\ 0 & & & & & \\ & & & b_n & a_n - \lambda \end{vmatrix} = 0.$$

The expansion of a general n x n determinant reduces to a system of n subdeterminants each (n-1) x (n-1). Each of the n subdeterminants will give rise to n-1 subdeterminants each (n-2) x (n-2). The tridiagonal form shown above is fortunate because only two of the subdeterminants are nonzero at each step. Thus the general determinant can be expressed as a sequence of polynomials:

$$f_m(\lambda) = (a_m - \lambda) f_{m-1}(\lambda) - b_m^2 f_{m-2}(\lambda).$$

If one lets

$$f_0(\lambda) = 1$$

and

$$f_1(\lambda) = a_1 - \lambda,$$

then for r=2,..., n one gets a sequence of polynomials known as a Sturm sequence. One property of this special type of sequence is that the roots of the $f_j(\lambda)$ polynomial separate the roots of the $f_{j+1}(\lambda)$ polynomial. Thus for $f_1(\lambda) = a_1 - \lambda = 0$, one can predict that the value $\lambda = a_1$ separates the two roots of $f_2(\lambda) = (a_2 - \lambda)(a_1 - \lambda) - b_2^2 = 0$. This information about the location of the roots of $f_2(\lambda)$ makes the iterative solution of this polynomial an easy process. Indeed, the technique of binary search can be utilized if one known bounds for the roots of the polynomial. As the process proceeds, the sequence leads to the solution of the final polynomial $f_n(\lambda) = 0$, which will give the n eigenvalues. This process can be illustrated by placing the roots of the sequence of polynomials as follows:

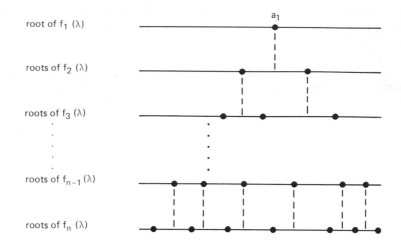

root of $f_1(\lambda)$

roots of $f_2(\lambda)$

roots of $f_3(\lambda)$

roots of $f_{n-1}(\lambda)$

roots of $f_n(\lambda)$

A further property of the Sturm sequence is that for any value of b such that $f_n(b) \neq 0$, the number of eigenvalues of A that are greater than b is equal to the number of sign changes in the sequence:

$$1, \; -f_1(b), f_2(b), \ldots \ldots (-1)^n f_n(b).$$

If this integer number of sign changes is called V(b), the number of eigenvalues between two real points b and c is V(b) - V(c).

3.6   Other Methods of Eigenvalue Calculation

In this section two significant methods for finding eigenvalues are discussed. The methods were both developed within the past 20 years and represent the most efficient methods presently available when all eigenvalues of a general real or complex matrix are desired. Both methods use transformations to produce a sequence of similar matrices which converge to a block triangular form:

In this form the X_m blocks represent 2 x 2 matrices on the diagonal. The eigenvalues of the X_m blocks are the eigenvalues of the original n x n matrix. This form is convenient because the second order determinant of the X_m blocks allows complex eigenvalues to be found without the need for complex terms in the final matrix. If all eigenvalues of the original matrix are pure real, the final form will be pure triangular with the eigenvalues on the diagonal.

The LR Method

The LR method was first developed by Rutishauser in 1958. In this method a matrix A is decomposed into:

$$A = LR,$$

where L is the unit lower triangular and R is upper triangular. Using the similarity transformation $L^{-1}AL$ we see that:

$$A_2 = L^{-1}AL = L^{-1}(LR)L = RL.$$

Thus:

and

$$A_{m-1} = L_{m-1}R_{m-1}$$

$$A_m = R_{m-1}L_{m-1}.$$

This process is repeated until L_s tends to the identity matrix I and R_s tends to the block diagonal form. Although this method is easy to apply, it may become unstable. For this reason the method discussed next is often preferred.

The QR Method

The QR method was first presented by Francis in 1961. The QR algorithm is defined by the relationship:

$$A_m = Q_m R_m$$

In this expression Q_m is an orthogonal matrix and R_m is an upper triangular matrix. As the method progresses:

$$A_{m+1} = Q_m^T A_m Q_m = Q_m^T Q_m R_m Q_m = R_m Q_m.$$

In the limit, the sequence of A matrices becomes the block diagonal form. This method is more difficult to implement and consumes more computer time than the LR method. Nevertheless, its stable numerical performance caused by the use of orthogonal transformation matrices causes it to be regarded as the best general solution method.

59

Example 3-3

Suppose that it is desired to find all eigenvalues of the general 6 x 6 matrix:

$$\begin{bmatrix} 2.3 & 4.3 & 5.6 & 3.2 & 1.4 & 2.2 \\ 1.4 & 2.4 & 5.7 & 8.4 & 3.4 & 5.2 \\ 2.5 & 6.5 & 4.2 & 7.1 & 4.7 & 9.3 \\ 3.8 & 5.7 & 2.9 & 1.6 & 2.5 & 7.9 \\ 2.4 & 5.4 & 3.7 & 6.2 & 3.9 & 1.8 \\ 1.8 & 1.7 & 3.9 & 4.6 & 5.7 & 5.9 \end{bmatrix}$$

The eigenvalues will be found by a two-step process in which similarity transformations will be used first to reduce the matrix to Hessenberg form and then a form of the QR method will be used to extract the eigenvalues. The following program which implements this solution, makes use of two subroutines from the IBM Scientific Subroutine Package. Subroutine HSBG reduces the 6 x 6 matrix to Hessenberg form, and the subroutine ATEIG then finds the eigenvalues.

```
C       ***************************************************
C       *       THIS PROGRAM FINDS ALL EIGENVALUES OF     *
C       *       A GENERAL 6 X 6 MATRIX. THE PROGRAM       *
C       *       USES SUBROUTINES HSBG AND ATEIG FROM      *
C       *       THE IBM SCIENTIFIC SUBROUTINE PACKAGE.    *
C       *               T. E. SHOUP 10/30/77              *
C       ***************************************************
        DIMENSION A(6,6),RR(6),RI(6),IANA(6)
        READ(5,100)((A(I,J),J=1,6),I=1,6)
        WRITE(6,104)
  104   FORMAT(///1X,'THE ORIGINAL MATRIX IS AS FOLLOWS')
        WRITE(6,103)
  103   FORMAT(1X,65('-'))
        WRITE(6,101)((A(I,J),J=1,6),I=1,6)
        WRITE(6,103)
  101   FORMAT(6(1X,F10.5))
  100   FORMAT(6F10.5)
        CALL HSBG(6,A,6)
        WRITE(6,105)
  105   FORMAT(///1X,'THE MATRIX IN HESSENBURG FORM IS')
        WRITE(6,103)
        WRITE(6,101)((A(I,J),J=1,6),I=1,6)
        WRITE(6,103)
        CALL ATEIG(6,A,RR,RI,IANA,6)
        WRITE(6,106)
  106   FORMAT(///1X,'THE EIGENVALUES ARE AS FOLLOWS')
        WRITE(6,107)

  107   FORMAT(1X,23('-'),/,4X,'REAL',12X,'IMAG',/,23('-'))
        WRITE(6,102)(RR(I),RI(I),I=1,6)
        WRITE(6,108)
  108   FORMAT(1X,23('-'))
  102   FORMAT(2(2X,F10.5))
        STOP
        END
```

The output of this program is:

```
THE ORIGINAL MATRIX IS AS FOLLOWS
-------------------------------------------------------------------
   2.30000   4.30000   5.60000   3.20000   1.40000   2.20000
   1.40000   2.40000   5.70000   8.40000   3.40000   5.20000
   2.50000   6.50000   4.20000   7.10000   4.70000   9.30000
   3.80000   5.70000   2.90000   1.60000   2.50000   7.90000
   2.40000   5.40000   3.70000   6.20000   3.90000   1.80000
   1.80000   1.70000   3.90000   4.60000   5.70000   5.90000
-------------------------------------------------------------------


THE MATRIX IN HESSENBURG FORM IS
-------------------------------------------------------------------
  -1.13162   3.20402  -0.05631   3.88246   1.40000   2.20000
  -0.75823   0.07468   0.48742   6.97388   5.37635  10.36283
   0.        1.13783  -2.63803  10.18618   7.15297  17.06242
   0.        0.        3.35891   7.50550   7.09754  13.92154
   0.        0.        0.       13.36279  10.58947  16.78421
   0.        0.        0.        0.        5.70000   5.90000
-------------------------------------------------------------------


THE EIGENVALUES ARE AS FOLLOWS
------------------------------
   REAL           IMAG
------------------------------
   25.52757      0.
   -5.63130      0.
    0.88433      3.44455
    0.88433     -3.44455
   -0.68247      1.56596
   -0.68247     -1.56596
------------------------------
```

3.7 Considerations in the Selection of an Eigenvalue Algorithm

The selection of an appropriate algorithm for a given eigenvalue problem will depend on the eigenvalue type, the matrix type, and the number of eigenvalues desired. As the complexity of the eigenvalue problem increases, the number of alternate algorithms decreases. To aid in the selection process, an application chart (Table 3-1) is presented. Most computer software packages contain subroutines that implement several or all of these algorithms. One creative way to apply this software is to use two of the subroutines together in order to make optimum use of their best characteristics. For example, if one has a general matrix, one could use Householder's method to reduce the matrix to Hessenburg form and then use the QR algorithm to find the eigenvalues. In this way the user has utilized both the speed advantages of Householder's method and the versatility of the QR algorithm.

Table 3-1   Applications Chart for Eigenvalue Algorithm Selection

Algorithm Name	Applied to	Result	Recommended when Number of Eigenvalues desired is			Comments
			largest or smallest	all ≤ 6	all ≥ 6	
Determinant (iteration)	General Matrix	eigen-values		*		Requires that the roots of a general polynomial be found.
Iteration (iteration)	General Matrix	eigen-values & eigen-vectors	*	*	*	Best accuracy for largest and smallest eigenvalues.
Jacobi (transformation)	Symmetric	diagonal form		*	*	Theoretically requires an infinite number of steps.
Given's (transformation)	Symmetric	tri-diagonal		*	*	Requires roots of an easy polynomial.
	Nonsymmetric	Hessenberg		*	*	Requires additional method.
Householder's (transformation)	Symmetric	tri-diagonal		*	*	Requires roots of an easy polynomial.
	Nonsymmetric	Hessenberg		*	*	Requires additional method.
LR method (transformation)	General Matrix	Block diagonal		*	*	Can be unstable.
QR Method (transformation)	General Matrix	Block diagonal		*	*	The best general method.

Problems

3.1 A triaxial stress tensor can be expressed as:

$$
\begin{bmatrix}
S_{xx} & \tau_{xy} & \tau_{xz} \\
\tau_{xy} & S_{yy} & \tau_{yz} \\
\tau_{xZ} & \tau_{yz} & S_{zz}
\end{bmatrix}
$$

Expand the eigenvalue determinant for this matrix to get a general cubic polynomial.

If: $S_{xx} = 30 \times 10^6$ N/m^2 $\tau_{xy} = 6 \times 10^6$ N/m^2

$S_{yy} = 40 \times 10^6$ N/m^2 $\tau_{yz} = 7 \times 10^6$ N/m^2

$S_{zz} = 20 \times 10^6$ N/m^2 $\tau_{xz} = 5 \times 10^6$ N/m^2

find the principal stress values using one of the root solving methods from Chapter 2.

3.2 For the principal stress eigenvalues found in Problem 3.1, find the corresponding eigenvectors.

3.3 Under what conditions would Newton's method of iteration not be successful in finding the eigenvalues in Problem 3.1?

3.4 By the method of iteration, find the largest and smallest eigenvalues for the matrix:

$$
\begin{bmatrix}
1 & 3 & 2 & 4 \\
5 & 9 & 4 & 1 \\
7 & 3 & 2 & 6 \\
8 & 7 & 8 & 4
\end{bmatrix}
$$

3.5 In dynamics a three-dimensional body will have three moments of inertia about three mutually perpendicular coordinate axes and three products of inertia about the three coordinate planes. For an unsymmetric body, it is found that for a given origin of coordinates there will be one orientation of the axes for which the products of inertia vanish. This orientation corresponds to the principal axes of inertia, and the corresponding moments of inertia about these axes are known as the principal moments of inertia. The principal moments of inertia include the maximum possible value, the minimum possible value, and an intermediate value. For the inertia matrix:

$$
\begin{bmatrix}
4.3 & 2.4 & 1.9 \\
2.4 & 3.2 & 2.7 \\
1.9 & 2.7 & 5.1
\end{bmatrix}
$$

find the three principal moments of inertia. What will be the rotation matrix Q that will produce the principal axes?

3.6 In the following matrix, all elements are constant except for the A(3,4) element:

$$\begin{bmatrix} 9.1 & 3.0 & 2.6 & 4.0 \\ 4.2 & 5.3 & 4.7 & 1.6 \\ 3.2 & 1.7 & 9.4 & X \\ 6.1 & 4.9 & 3.5 & 6.2 \end{bmatrix}$$

Find all eigenvalues for this matrix for X = 0.9, 1.0, and 1.1. From your results, can you say how much a 10% change in value for one element of a matrix will influence the resulting sizes of the eigenvalues?

3.7 Repeat Problem 3.6 using A(3,4) = 1.0 and A(3,3) = 8.46, 9.40, and 10.34. Are the eigenvalues in this matrix more sensitive to changes in the diagonal elements or the off-diagonal elements?

3.8 A barge is being designed to carry a string of six railroad cars across Lake Erie. The engine will be attached to a bulkhead as shown below. The car masses and coupling stiffness vary as shown. A concern has been raised about whether the string of cars might be longitudinally excited by wave motion. Calculate the six natural frequencies of the system shown, and compare them to the expected wave frequency of 1.0 radian per sec. The natural frequencies are related to the eigenvalues of the dynamic matrix D by:

$$\omega_i = \sqrt{1/\lambda_i}$$

The dynamic matrix is composed of the stiffness matrix [K] and the mass matrix [M]:

$$[D] = [K]^{-1}[M]$$

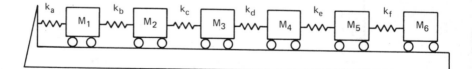

$$k_a = 5 \times 10^5 \text{ N/m}$$

$$k_b = k_c = k_d = k_e = k_f = 1 \times 10^5 \text{ N/m}$$

$$M_1 = 8 \times 10^4 \text{ kg}$$

$$M_2 = M_4 = 3 \times 10^4 \text{ kg}$$

$$M_3 = M_5 = 4 \times 10^4 \text{ kg}$$

$$M_6 = 2 \times 10^4 \text{ kg}$$

3.9 A cantilever beam 10 m long, with EI = 10^4 N/m² and mass of 10 kg/m is to be approximated by two point masses weighing 50 kg each. The masses are to be located at the center and at the free end as shown.

64

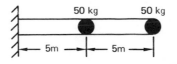

It is desired to find the two fundamental frequencies of vibration. These can be determined from the eigenvalues λ_i of the dynamic matrix $[D] = [F][M]$ where:

$$\omega_i = \sqrt{1/\lambda_i},$$

[M] is a diagonal matrix with the masses of each point along the major diagonal, and

[F] is the flexibility matrix in which the elements of the ith row are the deflections at points "j" due to a unit force at point "i."

There is no axial force, and shear deformations may be neglected.

3.10 Rework Problem 3.9 using five uniformly spaced masses rather than two, and compare the results.

References

1. Arden, Bruce W., and Astill, Kenneth N. Numerical Algorithms: Origins and Applications. Reading, Mass: Addison-Wesley Publ. Co., 1970.

2. Bathe, K. J., and Wilson, E. L. "Solution Methods for Eigenvalue Problems in Structural Mechanics." Intl. J. Num. Meth. Engr. 6 (1973): 213-26.

3. Berezin, I. S. and Zhidkov, N. P. Computing Methods. Vol. II, Oxford, England: Pergamon Press, 1965.

4. Gastines, Noel. Linear Numerical Analysis. New York: Academic Press, 1970.

5. Gupta, K. K. "Recent Advances in Numerical Analysis of Structural Eigenvalue Problems." Theory and Practice in Finite Element Structural Analysis, Edited by Y. Yamada and P. H. Gallagher, Proc. of the 1973 Tokyo Seminar on Finite Element Analysis, pp. 233-45.

6. Pilkey, Walter, and Pilkey, Barbara, eds. Shock and Vibration Computer Programs, Reviews and Summaries. The Shock and Vibration Information Center, United States Department of Defense, 1975.

7. Ralston, A. A First Course in Numerical Analysis. New York: McGraw-Hill Book Co., 1965.

8. Wilkinson, J. H. The Algebraic Eigenvalue Problem. London: Oxford University Press, 1965.

9. William, P. W. Numerical Computation. New York: Harper and Row, 1972.

4 COMPUTER SOLUTION OF ORDINARY DIFFERENTIAL EQUATIONS

This programmable energy controller contains a microprocessor that allows monitoring of consumption, prediction of energy demand, and regulation of power drawn to eliminate peaks. [Photo courtesy of Westinghouse Electric Corp.]

4.1   Introduction

Any equation containing one or more derivatives is called a differential equation. Since most physical laws of engineering are expressed in terms of differential equations, the need to solve this type of equation occurs again and again in the modeling and evaluation phases of the design process. Indeed, any design-analysis problem concerned with the movement of mass or energy will ultimately lead to a differential equation. Unfortunately, the number of differential equations that can be satisfactorily treated by hand calculations is quite limited. For this reason the topic of computer solution of differential equations is especially important to a discussion of engineering problem solving.

Differential equations fall into two distinct categories depending on the number of independent variables they contain and thus on the type of derivatives they contain. If only one independent variable exists, the derivatives will be ordinary, and the equation will be called an "ordinary differential equation." If more than one independent variable exists, the derivatives must be partial, and the equation will be called a "partial differential equation." The purpose of this chapter is to discuss solution methods for ordinary differential equations. The topic of partial differential equations is considered separately in a later chapter.

Throughout this chapter it is assumed that the reader has a basic understanding of the terminology of differential equations. The reader who feels deficient in this area may wish to undertake a brief review before proceeding.

4.2   Categories of
Ordinary Differential Equations

The solution of problems involving differential equations requires that the dependent variable and/or its derivatives be given at prescribed values of the independent variable. Whenever all of these constraints are imposed at the zero value of the independent variable, the solution task is said to be an "initial-value problem." Whenever any of the constraints are imposed at any independent variable vlaue other than zero, the solution task is said to be a "boundary-value problem." Constraints on initial-value problems are generally termed "initial conditions," and the constraints on boundary-value problems are termed "boundary conditions." The most common type of independent variable used in an initial-value problem is time. For example, the free vibration of a spring-supported mass would have its displacement described by a differential equation using time as the independent variable. If the constraints on the motion are given for the displacement and velocity at time equal to zero, the problem would be categorized as an initial-value problem. This same device would give rise to a boundary-value problem if one of the constraints involved a given displacement at the end of a prescribed time interval. Frequently, a boundary-value problem has length as its independent variable rather than time. A familiar example of this situation is the differential equation that describes the deflection of a beam. In this situation boundary conditions are usually specified at the two ends. Although these two categories of problems are being discussed in the same chapter, the computational differences in their numerical solutions are substantial. Both cateogries are considered in this chapter. Initial-value problems are considered first.

Initial-Value Problems

The initial-value problem can be stated in simple terms. Given an initial condition $y(x_0) = y_0$ and the differential equation:

$$\frac{dy}{dx} = f(x, y),$$

find the unknown function $y(x)$ that satisfies both the differential equation and the initial condition. Generally speaking, a numerical solution to this problem is accomplished by first computing the slope of the curve and then proceeding in small increments of x to a new point $x_1 = x_0 + h$. The new point is obtained using information about the slope of the curve as calculated from the differential equation. Thus the numerical solution is composed of a series of short, straight-line segments that approximate the true curve of $y = f(x)$. The numerical method itself is concerned with the calculation procedure for moving from point to point on the curve.

Because the topic of numerical solution of initial-value problems is of such importance to many fields of science and engineering, this topic has received considerable attention over the years. As a result, an extremely large number of methods exist. No attempt is made here to discuss every such algorithm; rather, the most common of those that represent the following two categories of initial-value problems are discussed.

1. One step methods are those methods that use information about a single, previous step to find the next point on a curve $y = f(x)$. Methods of this type include Euler's method and the Runge-Kutta methods.

2. Predictor-corrector methods, also known as multistep methods, find the next point on a curve $y = f(x)$ by using information from more than one previous point. Iteration is often used to achieve a sufficiently close numerical value. Methods of this type include Milne's method, the Adams-Bashforth method, and the Hamming method.

Errors

Before discussing the specific differential equation methods, it is important to understand the source of errors associated with the numerical approximations. There are three sources of errors that occur:

1. Round-off error is due to the numerical limitations of the computer being used. Every computer has a limitation on the number of significant digits that it can store and manipulate.

2. Truncation error is due to the fact that infinite series used to approximate a function are often truncated after a few terms. Such a procedure is commonplace in numerical methods and introduces errors that are due entirely to the method rather than the machine.

3. Propagation error is due to the accumulation of previous errors in a numerical scheme. Any approximate technique is never exact. Thus any error that an approximation scheme introduces at an early step will be carried along in the computation process for later steps.

These three sources of errors give rise to two types of observed errors.

1. Local error is the amount of error that enters the computational process at any given computational step.

2. Global error at any point in the computation is the difference between the computed value of the solution and the exact solution. Thus the global error accounts for the total accumulation of error from the start of the computational process.

4.3 One-Step Methods for the Initial-Value Problem

The one-step methods may be used to solve a first order differential equation of the form:

$$y' = f(x, y),$$

where y' represents dy/dx. The initial condition for this problem is expressed as $y(x_0) = y_0$. The purpose of a one-step method is to provide a means for calculating a sequence of y values corresponding to discrete values of the independent variable.

Euler's Method

Euler's method is the simplest of all methods for solving the initial value problem. It is applicable to first-order differential equations and has rather limited accuracy. Because of this characteristic, it is not recommended for practical use. It does, however, provide considerable insight into the understanding of several other, more practical methods.

The basis for the Euler method comes from an application of the initial conditions to a Taylor's series of the form:

$$y(x_0 + h) = y(x_0) + hy'(x_0) + \tfrac{1}{2}h^2y''(x_0) + \ldots .$$

If the value of h is quite small, terms containing powers of h^2 and higher will be much smaller and will be neglected. When this is done, the expansion becomes:

$$y(x_0 + h) = y(x_0) + hy'(x_0).$$

Evaluation of the differential equation at the initial condition will give $y'(x_0)$. Thus the dependent variable can be approximated at a small step "h" away from the initial point. This process can be continued for as many steps as desired using the relationship:

$$y_{n+1} = y_n + hf(x_n, y_n) \qquad n = 1, 2, \ldots .$$

The Euler method is presented in graphical form in Figure 4-1. The error

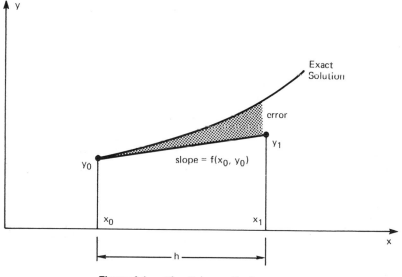

Figure 4-1   The Euler method.

in this method is said to be on the order of "h^2," because terms containing "h^2" and higher were neglected in the original Taylor's expansion.

Modified Euler Method

One of the difficulties associated with the use of Euler's method is that, although the slope of the exact curve at the starting point is $y'(x_0)$, this slope changes as the independent variable changes. Thus at the point $x_0 + h$, the slope is no longer the same as it was at the start of the interval. Thus an error is introduced into the calculation process whenever the starting slope is used for the whole interval. The accuracy of the Euler method can be substantially improved if a better approximation is used for the derivative. One possible way to do this is to use an average value of the derivatives at the beginning and end of the interval. The modified Euler method does this by taking a temporary Euler step:

$$y_{n+1}^* = y_n + hf(x_n, y_n)$$

and then using y_{n+1}^* to calculate an approximation to the derivative at the end of the interval. This derivative will be $f(x_{n+1}, y_{n+1}^*)$. This new derivative is averaged with the initial derivative to obtain a more accurate value for y_{n+1}:

$$y_{n+1} = y_n + \tfrac{1}{2}h \left[f(x_n, y_n) + f(x_{n+1}, y_{n+1}^*) \right].$$

This procedure is illustrated in Figure 4-2. Another way to understand

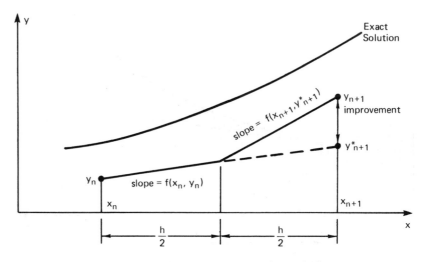

Figure 4-2 The modified Euler method.

how the modified Euler method is developed is to return to the Taylor's expansion:

$$y(x_0 + h) = y(x_0) + hy'(x_0) + \frac{h^2}{2} y''(x_0) + \ldots..$$

If the h^2 terms are retained and all higher-order terms are dropped, it seems reasonable to expect more accuracy. It should be noted, however, that saving the h^2 term requires that the second derivative $y''(x_0)$ be known. This second derivative can be approximated by a finite difference:

$$y''(x_0) = \frac{\Delta y'}{\Delta x} = \frac{y'(x_0 + h) - y'(x_0)}{h}$$

If this expression is substituted into the truncated Taylor's expansion, the result is:

$$y(x_0 + h) = y(x_0) + \tfrac{1}{2}h \ [y'(x_0 + h) + y'(x_0)]$$

which is equivalent to the expression found previously.

This method is called a "second-order method" because it utilizes the h^2 terms in the Taylor's expansion. For this method the error at each step is said to be on the order of h^3. The price one pays to achieve this improved accuracy is the time spent in the extra calculation to get y^*_{n+1}. This method suggests that additional accuracy could be achieved if the user were willing to invest the time to achieve an even better approximation for the derivative (i.e., if the user were willing to include additional terms in the Taylor's expansion). This concept is the basis for the Runge-Kutta methods.

Runge-Kutta Methods

In order to retain the nth-order term in a Taylor's expansion, it is necessary to be able to evaluate the nth-order derivative of the dependent variable. In the modified Euler method, the slopes at the two ends of the interval of interest were sufficient to provide a finite difference form for the second derivative. In order to use a finite difference form to calculate the third derivative, it will be necessary to know the second derivative at at least two different locations. This requires that an additional slope be evaluated at an intermediate point within the "h" interval from x_n to x_{n+1}. Indeed, the higher the order of the derivative that must be calculated, the higher will be the number of additional, internal evaluations that must be made. The Runge-Kutta method provides a set of formulas for selecting the spacing of the internal evaluations required to implement this strategy. Since a number of alternatives exist for the spacing and for the relative weighting to be used for the slopes found, the term "Runge-Kutta" refers to a large family of methods for handling first-order differential equations. The most commonly used Runge-Kutta formulation is based on retaining all terms up through "h^4" and is thus termed a "fourth-order" method having an error at any step on the order of h^5. The calculation formula for this classical method is:

$$y_{n+1} = y_n + \frac{(K_0 + 2K_1 + 2K_2 + K_3)}{6}$$

where:

$$K_0 = hf(x_n, y_n)$$

$$K_1 = hf(x_n + 0.5h, y_n + 0.5 K_0)$$

$$K_2 = hf(x_n + 0.5h, y_n + 0.5 K_1)$$

and

$$K_3 = hf(x_n + h, y_n + K_2).$$

In reality the Euler and modified Euler methods are actually first- and second-order Runge-Kutta methods. The increased accuracy made possible by the Runge-Kutta technique makes it far more desirable to use than either of the previous two methods discussed. The increased accuracy of the Runge-Kutta method more than justifies the additional computational effort needed to use it. Because of this greater accuracy, it is often possible to use a larger sized "h" interval. The allowable error at each step will determine the maximum allowable step size that can be used. It is wise to adjust the value of "h" to the maximum allowable in order to achieve optimum efficiency in the computational process. Frequently, this adjustment process is included as an automatic part of a computational scheme employing the Runge-Kutta method.

The best way to illustrate the relative accuracy of the one-step methods is by means of an example.

Example 4-1

Suppose that it is desired to find the solution to the equation:

$$\frac{dy}{dx} = 2x^2 + 2y$$

subject to $y(0) = 1.$, on the interval $0 \leq x \leq 1.$, with $h = 0.1$. Because this problem is linear, the exact solution is known to be:

$$y = 1.5e^{2x} - x^2 - x - 0.5$$

and can be used to inspect the relative accuracy of the various methods. A comparison of the results is presented in tabular form below. Clearly, the Runge-Kutta method is better than the Euler method or the modified Euler method.

x_n	Euler	Modified Euler	Runge-Kutta	Exact
0.0	1.0000	1.0000	1.0000	1.0000
0.1	1.2000	1.2210	1.2221	1.2221
0.2	1.4420	1.4923	1.4977	1.4977
0.3	1.7384	1.8284	1.8432	1.8432
0.4	2.1041	2.2466	2.2783	2.2783
0.5	2.5569	2.7680	2.8274	2.8274
0.6	3.1183	3.4176	3.5201	3.5202
0.7	3.8139	4.2257	4.3927	4.3928
0.8	4.6747	5.2288	5.4894	5.4895
0.9	5.7376	6.4704	6.8643	6.8645
1.0	7.0472	8.0032	8.5834	8.5836

Runge-Kutta Methods for a System of Differential Equations

Any of the Runge-Kutta formulas can be used to solve simultaneous differential equations and thus higher-order differential equations since one can make "n" first-order differential equations from a single n^{th}-order differential equation. For example, in the second order differential equation:

$$\frac{d^2y}{dx^2} = g(x, y, \frac{dy}{dx}),$$

one can let

$$z = \frac{dy}{dx}$$

and then

$$\frac{dz}{dx} = \frac{d^2y}{dx^2}.$$

The two first-order equations become:

$$\frac{dz}{dx} = g(x, y, z) \quad \text{and} \quad \frac{dy}{dx} = f(x, y, z),$$

where in this case $f(x, y, z) = z$. The initial-value problem for this situation would be specified in terms of two initial conditions:

$$y(x_0) = y_0 \text{ and } z(x_0) = z_0.$$

The Runge-Kutta formulas for this problem would be:

$$y_{n+1} = y_n + K$$

and

$$z_{n+1} = z_n + L,$$

where

$$K = \frac{(K_1 + 2K_2 + 2K_3 + K_4)}{6}$$

and

$$L = \frac{(L_1 + 2L_2 + 2L_3 + L_4)}{6}$$

In these equations:

$$K_1 = hf(x_n, y_n, z_n)$$
$$L_1 = hg(x_n, y_n, z_n)$$
$$K_2 = hf(x_n + 0.5h, y_n + 0.5K_1, z_n + 0.5L_1)$$
$$L_2 = hg(x_n + 0.5h, y_n + 0.5K_1, z_n + 0.5L_1)$$

75

$$K_3 = hf(x_n + 0.5h, \ y_n + 0.5K_2, \ z_n + 0.5L_2)$$

$$L_3 = hg(x_n + 0.5h, \ y_n + 0.5K_2, \ z_n + 0.5L_2)$$

$$K_4 = hf(x_n + h, \ y_n + K_3, \ z_n + L_3)$$

$$L_4 = hg(x_n + h, \ y_n + K_3, \ z_n + L_3)$$

A Summary of One-Step Methods

Certain special characteristics are common to all one-step methods.

1. They need information only at the preceding point to find information at the new point. This characteristic is known as "self-starting behavior."

2. They all stem from a Taylor's expansion of the function and contain terms up to and including the one containing h to the power k. The integer "k" is called the "order" of the method, and the error at any step is said to be on the order of k + 1.

3. They do not require the actual evaluation of any derivatives but only of the function itself. They may, however, require several evaluations of the function at intermediate points. This, of course, can be time- and effort-consuming.

4. Their self-starting nature lends itself to easy changes in the step size "h".

Example 4-2

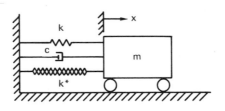

Shock and vibration problems in the aerospace and transportation industries arise from many different types of excitation sources. The elimination of shock and vibration is of crucial importance in the isolation of instruments and controls or in the protection of human occupants of vehicles. The usual solution to problems involving excess vibration transmission involves the use of lightly damped flexible supports. These soft supports cause the natural frequency of a suspension system to be far below the disturbing frequency. This solution is effective for the isolation of steady-state vibration; however, when these suspensions encounter shock excitation, their softness often leads to damagingly large deflections. It has been pointed out that this undesirable feature is not present in suspension systems utilizing symmetrically nonlinear springs that harden.† These springs become progressively stiffer when subjected to large deflections from the "operating point." The device shown consists of a mass m connected to the rigid wall by means of a linear spring with coefficient k, a damper with coefficient c, and a nonlinear spring that exerts a restoring force proportional to a constant k* times the third power of displacement. This "cubic" spring will provide a symmetrically nonlinear behavior to satisfy the need for shock and vibration isolation.

Because the resulting differential equation for the motion of this sytem is described by the nonlinear differential equation:

$$m\ddot{x} + c\dot{x} + kx + k*x^3 = 0,$$

the displacement x as a function of time cannot be found by traditional "exact" methods. For this reason a numerical solution to this differential equation is desirable.

If the physical parameters of the suspension system are as follows:

$$k = 2.0 \text{ N/cm}$$

$$k* = 2.0 \text{ N/cm}^3$$

$$c = 0.15 \text{ N sec/cm}$$

$$m = 1.0 \text{ kg}$$

and the initial conditions are:

$$x(0) = 10. \text{ cm}$$

$$\dot{x}(0) = 0. \text{ cm/sec,}$$

prepare and run a computer program that will simulate the motion of this system for time from 0 to 1.0 seconds.

In order to solve this problem by a one-step method it will be necessary to reduce the second-order differential equation to two first-order differential equations. When this is done, the result will be:

$$\dot{x}_1 = x_2$$

$$\dot{x}_2 = \frac{-c}{m}\dot{x}_1 - \frac{k}{m}x_1 - \frac{k*}{m}x_1^3.$$

In order to utilize centimeter dimensions in the output, it is necessary to convert the mass to

$$m = 0.01 \text{ N sec}^2/\text{cm.}$$

A FORTRAN program that solves this problem utilizes a Runge-Kutta method using a subroutine RKGS, which is part of the IBM Scientific Subroutine Package [16]. This subroutine utilizes two external subprograms in accomplishing its task. The first subroutine FCT provides the two first-order differential

---

†Tobias, J. A., "Design of Small Isolator Units for the Suppression of Low-Frequency Vibration," _Journal of Mechanical Engineering Science_, Vol. 1, No. 3, 1959, pp. 280-292.

equations. The second subroutine OUTP provides a printout of the differential equation solution at intermediate values of time. Utilizing subroutine RKGS requires that the user supply an upper bound on the errors. Once this is done, the subroutine automatically adjusts the computational increment during the whole computation by halving or doubling a predetermined starting size. An observation of the printout provided by this computer program illustrates how the time increment is adjusted in order to maintain the desired error in an efficient fashion.

```
C     ****************************************************
C     *    THIS PROGRAM PROVIDES A NUMERICAL SOLUTION    *
C     *    TO THE VIBRATION OF A SPRING, MASS, DAMPER     *
C     *    SYSTEM CONTAINING A NONLINEAR SPRING.          *
C     *                         T. E. SHOUP  8/23/77     *
C     ****************************************************
      EXTERNAL FCT,OUTP
      DIMENSION P(5), X(2), XDOT(2), AUX(8,2)
C     INITIAL VALUE OF INDEPENDENT VARIABLE.
      P(1) = 0.
C     FINAL VALUE OF INDEPENDENT VARIABLE.
      P(2) = 1.0
C
C     INCREMENT SIZE
      P(3) = 0.02
C
C     UPPER BOUND ON ERRORS
      P(4) = 0.001
C
C     ERROR WEIGHTS (SUM MUST EQUAL 1.)
C
      XDOT(1) = 0.5
      XDOT(2) = 0.5
C
C     INITIAL VALUES
C
      X(1) = 10.0
      X(2) = 0.0
C
C     WRITE HEADING ON PAGE.
      WRITE (6,100)
  100 FORMAT (1H1,/,36('-'),/,1X,'TIME(SEC)',9X,'X(CM)',2X,
     &'X''(CM/SEC)',/,36('-'))
      CALL RKGS(P, X, XDOT, 2, IHLF, FCT, OUTP, AUX)
      WRITE(6,101)
  101 FORMAT (36('-'))
      STOP
      END
      SUBROUTINE FCT(T, X, XDOT)
      DIMENSION X(2), XDOT(2)
      XM = 0.01
      C = 0.15
      XK = 2.0
      XKS = 0.2
      XDOT(1) = X(2)
      XDOT(2) = -C*XDOT(1)/XM-XK*X(1)/XM-XKS*X(1)**3/XM
      RETURN
      END
      SUBROUTINE OUTP(T, X, XDOT, IHLF, NDIM, P)
      DIMENSION P(5), X(2), T(1), XDOT(2)
      WRITE (6,100) T(1), X(1), X(2)
  100 FORMAT (3(2X,F10.5))
      RETURN
      END
```

The output of this computer program is:

TIME(SEC)	X(CM)	X'(CM/SEC)
0.	10.00000	0.
0.00250	9.93232	-53.63445
0.00500	9.73515	-103.30084
0.00750	9.42045	-147.44249
0.01000	9.00345	-185.00307
0.01250	8.50137	-215.47514
0.01500	7.93200	-238.86480
0.01750	7.31261	-255.59608
0.02000	6.65898	-266.38511
0.02250	5.98491	-272.11187
0.02500	5.30187	-273.70900
0.02750	4.61905	-272.07729
0.03000	3.94349	-268.02979
0.03250	3.28033	-262.26127
0.03500	2.63314	-255.33710
0.03750	2.00424	-247.69558
0.04000	1.39499	-239.65846
0.04500	0.23779	-223.19034
0.05000	-0.83702	-206.75019
0.05500	-1.82965	-190.24708
0.06000	-2.73839	-173.07487
0.06500	-3.55794	-154.42987
0.07000	-4.27900	-133.57957
0.07500	-4.88931	-110.09365
0.08000	-5.37558	-84.02055
0.08500	-5.72618	-55.97076
0.09000	-5.93384	-27.06928
0.09500	-5.99788	1.22851
0.10000	-5.92511	27.41301
0.10500	-5.72945	50.20470
0.11000	-5.43018	68.75991
0.11500	-5.04951	82.74876
0.12000	-4.61011	92.30897
0.12500	-4.13303	97.91907
0.13000	-3.63640	100.24597
0.13500	-3.13484	100.00972
0.14000	-2.63945	97.88706
0.14500	-2.15815	94.45665
0.15000	-1.69629	90.17825
0.15500	-1.25720	85.39456
0.16000	-0.84278	80.34567
0.17000	-0.09090	70.01650
0.18000	0.55797	59.79153
0.19000	1.10562	49.76552
0.20000	1.55374	39.87560
0.21000	1.90349	30.09752
0.22000	2.15638	20.52929
0.23000	2.31552	11.39368
0.24000	2.38670	2.98679
0.25000	2.37869	-4.39720
0.26000	2.30293	-10.53539
0.27000	2.17253	-15.31771
0.28000	2.00109	-18.75349
0.29000	1.80161	-20.95007
0.30000	1.58567	-22.07721
0.31000	1.36298	-22.33086
0.32000	1.14131	-21.90480
0.34000	0.72304	-19.68125
0.36000	0.36060	-16.45956
0.38000	0.06689	-12.89064

0.40000	-0.15532	-9.35702
0.42000	-0.30923	-6.09170
0.44000	-0.40181	-3.24308
0.46000	-0.44236	-0.89990
0.48000	-0.44145	0.90124
0.50000	-0.40988	2.17024
0.52000	-0.35789	2.95410
0.54000	-0.29448	3.32438
0.56000	-0.22712	3.36460
0.58000	-0.16153	3.15982
0.60000	-0.10182	2.78943
0.62000	-0.05059	2.32288
0.64000	-0.00916	1.81760
0.66000	0.02216	1.31847
0.68000	0.04384	0.85816
0.70000	0.05690	0.45819
0.72000	0.06266	0.13039
0.74000	0.06262	-0.12141
0.76000	0.05829	-0.29961
0.78000	0.05108	-0.41104
0.80000	0.04224	-0.46534
0.82000	0.03278	-0.47360
0.84000	0.02352	-0.44719
0.86000	0.01505	-0.39692
0.88000	0.00774	-0.33242
0.90000	0.00179	-0.26182
0.92000	-0.00274	-0.19153
0.94000	-0.00590	-0.12631
0.96000	-0.00784	-0.06932
0.98000	-0.00874	-0.02234
1.00000	-0.00881	0.01400

From the following plot of the displacement portion of this data, it is obvious that the frequency of vibration of the system is dependent on the amplitude of the vibration. This behavior is characteristic of nonlinear systems that contain springs that harden as they deflect.

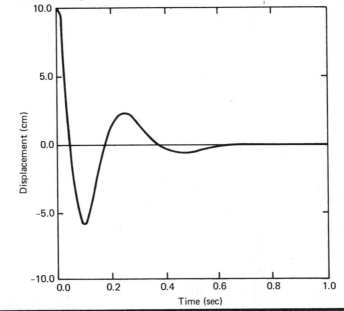

4.4   Predictor-Corrector Methods

Predictor-corrector methods use information from several previous points to compute a new point. They make use of two formulas--one called a predictor equation and the other called a corrector equation. Regardless of the predictor-corrector method used, the logic flow will be the same. Only the prediction and correction formulas are changed to change methods. The logic of this differential equation technique is presented in Figure 4-3 for the solution of the differential equation:

$$y'(x) = f(x, y).$$

Since the predictor-corrector methods require information about several previous points in order to move forward, they are not self-starting as were the one-step methods. For this reason all predictor-corrector methods require the use of a one-step method to get started. A Runge-Kutta method is frequently employed to meet this need. The method then proceeds as follows. First, the prediction formula is applied to the starting information to predict a value for $y_{n+1}^{(0)}$. The superscript (0) indicates that this prediction is just one of a sequence of y_{n+1} values that are better and better in terms of accuracy. Using this initial prediction for y_{n+1}, the differential equation is employed to compute the derivative:

$$y_{n+1}^{(0)'} = f(x_{n+1}, y_{n+1}^{(0)}).$$

Once this derivative is available, it is used in the corrector formula to obtain an improved value $y_{n+1}^{(j+1)}$. This value is, in turn, used to improve the derivative by means of the differential equation:

$$y_{n+1}^{(j+1)'} = f(x_{n+1}, y_{n+1}^{(j+1)}).$$

If this derivative is not sufficiently close to the previous one, the new value is applied to the corrector formula to continue the iterative improvement process. If, on the other hand, the change in derivatives is small enough, the value of $y_{n+1}^{(j+1)'}$ is used to compute and to document the final value of y_{n+1}. Once this has been achieved, the process may be repeated to move onward to the next step point y_{n+2}.

The usual approach to the derivation of predictor or corrector formulas is to view the stepping problem as if it were the process of approximate integration and thus to use finite-difference methods to generate the formulas.

If the differential equation $y' = f(x, y)$ is integrated between the limits x_n and x_{n+k}, the result is:

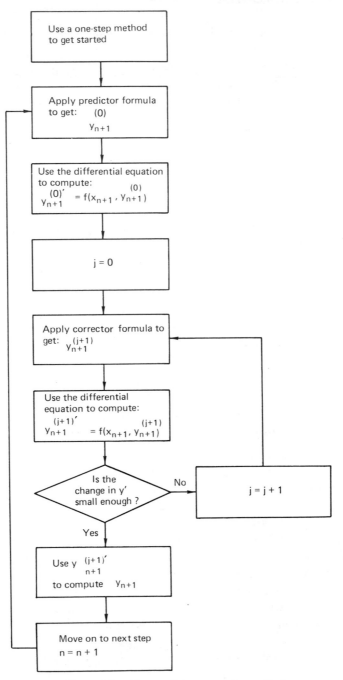

Figure 4-3 The predictor-corrector method.

$$y(x_{n+k}) - y(x_n) = \int_{x_n}^{x_{n+k}} f(x, y)dx.$$

This integral cannot be found directly because the relationship $y'(x)$ is not known in advance. A number of finite-difference methods to approximate this integral do exist. The particular selection will characterize the individual method being used. Any numerical integration formula that does not require a prior estimate of $y'(x_{n+1})$ is suitable for use as a predictor. Any numerical integration formula that does require a prior estimate of $y'(x_{n+1})$ is suitable for use as a corrector.

Milne's Predictor-Corrector Method
Milne's method [12] uses Milne's formula:

$$y_{n+1} = y_{n-3} + \frac{4h}{3} [2y'_n - y'_{n-1} + 2y'_{n-2}] + \frac{28}{90} h^5 y^{(5)}$$

as a predictor and Simpson's rule:

$$y_{n+1} = y_{n-1} + \frac{h}{3} [y'_n + 4y'_n + y'_{n-1}] - \frac{1}{90} h^5 y^{(5)}$$

as a corrector. The final terms in each of these formulas are not actually used in the iteration process; rather, they are included because they give an indication of the truncation error. Milne's method is termed a fourth order method because the truncation error is on the order of h^5. The user may be tempted to ask why it is necessary to bother with the corrector at all if the predictor is of fourth order. The answer to this is revealed by the relative size of the two error terms. In this case the corrector truncation error is 28 times smaller and is thus much more desirable. In general, iterative formulas are substantially more accurate than forward formulas. For this reason they are worthwhile to use in spite of their added difficulty. Although Milne's formulation has a small numerical coefficient in its error remainder (1/90), it is less often used than other methods with poorer remainders owing to the fact that it is inherently unstable. This means that the propagated error may grow exponentially. This characteristic is true for all corrector formulas based on Simpson's rule.

Adams-Bashforth Method
The Adams-Bashforth method [12] is a fourth-order method that has its predictor equation based on an integration of Newton's backward interpolation formula. For this method the predictor formula is:

$$y_{n+1} = y_n + \frac{h}{24} [55y'_n - 59y'_{n-1} + 37y'_{n-2} - 9y'_{n-3}] + \frac{251}{720} h^5 y^{(5)}$$

and the corrector formula is

$$y_{n+1} = y_n + \frac{h}{24} [9y'_{n+1} - 19y'_n - 5y'_{n-1} + y'_{n-2}] - \frac{19}{720} h^5 y^{(5)}$$

This method proceeds just as does Milne's method, but an error introduced at one stage in the Adams-Bashforth procedure does not tend to grow exponentially.

One might reasonably suspect that since the estimate of the error term is known, it could be used to improve the corrected value. Although this could be done, Ralston [25] comments that this process is equivalent to using a system of one higher order. Since the process of correcting the corrector can adversely influence the stability of the corrector, it is a better idea to use a higher-order formulation if higher accuracy is desired.

Hamming's Method
Hamming's method [14] is based on the following computational formulas:

Predictor: $\qquad y_{n+1}^{(0)} = y_{n-3} + \frac{4h}{3} (2y'_n - y'_{n-1} + 2y'_{n-2})$

Modifier: $\qquad \bar{y}_{n+1}^{(0)} = y_{n+1}^{(0)} + \frac{112}{121} [y_n - y_n^{(0)}]$

$$[\bar{y}_{n+1}^{(0)}]' = f(x_{n+1}, \bar{y}_{n+1}^{(0)})$$

Corrector: $\qquad y_{n+1}^{(j+1)} = \frac{1}{8}(9y_n - y_{n-2})$

$$+ \frac{3h}{8} \{[\bar{y}_{n+1}^{(j)}]' + 2y'_n - y'_{n-1}\}$$

This stable, fourth-order method is based on the predictor equation:

$$y_{n+1} = y_{n-3} + \frac{4h}{3} [2y'_n - y'_{n-1} + 2y'_{n-2}] + \frac{28}{90} h^5 y^{(5)}$$

and a corrector equation

$$y_{n+1} = \frac{1}{8} [9y_n - y_{n-2} + 3h(y'_{n+1} + 2y'_n - y'_{n-1})] - \frac{1}{40} h^5 y^{(5)}$$

The method has an added feature that uses estimates of the errors in the predictor and corrector to "mop up" their errors. Of all of the predictor-corrector methods, Hamming's method is one of the most often used because of its simplicity and stability.

Summary of Predictor-Corrector Characteristics
When compared with one-step methods, predictor-corrector methods have certain important characteristics.

1.   Predictor-corrector methods require information about prior points and therefore are not self-starting. Indeed, they must rely on some type of one-step method to get their start. If a change in step size is made during the solution process, a temporary reversion to the one-step starter method is usually required.

2.   Because the predictor-corrector methods need information about prior points, they also require the computer capacity to store this information.

3.   The one-step methods are of comparable accuracy to the predictor-corrector methods of the same order. The predictor-corrector methods provide an easy measure of the per-step error, while the one step methods do not. For this reason the step size "h" is often chosen conservatively smaller for the one-step methods than would otherwise be necessary. This tends to make the predictor-corrector methods appear to be more efficient.

4.   The actual number of functional evaluations for each step of a fourth order Runge-Kutta method is four, while a predictor-corrector method of the same order often requires only two evaluations to achieve convergence. For this reason the predictor-corrector methods can be almost twice as fast as the Runge-Kutta methods of comparable accuracy. This time saving can become a significant consideration in the selection of an algorithm because computer time can be expensive.

4.5   Step Size Considerations

One of the significant practical problems faced by the engineering programmer who is solving a differential equation is that of selecting a suitable step size for the computational process. If the step size is too small, the computational process will consume needless computer time, and the number of "per-step" errors contributing to the global error will be significant. If, on the other hand, the step size is too large, the local error due to truncation will be significant, and the resulting accumulation of global errors will cause the computational results to be of poor accuracy.

The most common procedure used in selecting the step size is to keep the local per-step error below a predetermined, allowable value. In general, for a method of order "n" the local error will be on the order of a constant times the step size raised to the power "n + 1." This may be expressed as:

$$C \, h^{n+1}$$

If the method being used is a predictor-corrector method, the per-step error is often presented as the last term in the corrector formula (see, for example, the discussion of Milne's method). If the Runge-Kutta method is being used, however, the local error is not so obvious. One way to estimate this error is based on Richardson [27] extrapolation which is discussed in the following paragraphs .

If the step size h is used to predict a value y_{j+1} of the solution at the point x_{j+1}, then the difference between this value and the true value y_{true} will be:

$$y_{true} - y_{j+1} = C\, h^{n+1}.$$

If a step size of h/2 is used to predict the value y^*_{j+1} at x_{j+1}' then the difference between this new value and the true value will be:

$$y_{true} - y^*_{j+1} = C \left(\frac{h}{2}\right)^{n+1}.$$

If this equation is subtracted from the previous expression, the result will be:

$$y_{j+1} - y^*_{j+1} = C\, h^{n+1} - C \left(\frac{h}{2}\right)^{n+1}$$

$$= C\, h^{n+1}[1 - (\tfrac{1}{2})^{n+1}].$$

Thus one can solve for the estimate of the local error:

$$C\, h^{n+1} = \frac{(y_{j+1} - y^*_{j+1})(2^{n+1})}{2^{n+1} - 1}.$$

The disadvantage to this method is that it requires the user to compute the value of y_{j+1} twice. Since the y^*_{j+1} computation requires the two half-size steps to get the value at x_{j+1}, the computational effort is more than doubled. Nevertheless, this procedure is often included in the computational algorithm to make automatic adjustments in the step size as the computational process takes place. This approach is frequently used for Runge-Kutta routines. Alternatively, if the per-step error is too large for a given step size, the error could be reduced by utilizing a higher order term in the computational process. This is, of course, most easily accomplished for the predictor-corrector methods.

The chief advantage of the Runge-Kutta methods are their ability to start easily and the ease with which the step sizes can be changed during the computational process. On the other hand, the primary advantage of the predictor-corrector methods is the ease with which the per-step error can be estimated. Although in the past these advantages were often regarded as mutually exclusive, recent powerful techniques allow the user to exploit the computational advantages of both types of methods. Such hybrid methods can be quite useful in the solution of engineering problems.

As an example, Gear [8] presents a method that features automatic control of both the computational step size and the degree of the predictor-corrector formulas.

4.6   Stiff Problems

Some types of ordinary differential equation problems do not lend themselves to solution by any of the methods previously discussed. To understand why this is so requires an understanding of the component parts to the solution of a differential equation. The time constant of a first-order differential equation is the time required for the transient portion of the solution to decay by a factor of $1/e$. A differential equation of nth-order will in general have n time constants. If any two of these time constants are widely different in magnitude or if one of the time constants is quite small relative to the solution interval, the problem is said to be "stiff" and will perform poorly when treated by traditional methods. Such systems require that the solution technique have small enough step sizes to account for the fastest component part of the process even after the contribution of this part has died out. Failure to maintain a small enough step size will lead to instability in the solution process. Although the difficulty in maintaining stability when using traditional methods on "stiff" systems can be temporarily averted by using small step sizes, this approach has two disadvantages. First, the use of extremely small step sizes relative to the size of the solution interval will cause the method to consume considerable time in achieving a complete solution. Second, the round-off and truncation errors that are amplified and accumulated through the use of many calculations will eventually lead to meaningless results.

Because "stiff" problems can occur in process control problems, electronic network problems, and chemical reaction problems, a number of recent research efforts have been directed toward discovering efficient methods for treating such problems. Although a discussion of these methods is beyond the scope of this text, the interested reader is encouraged to consult the works of Gear [7] and Hall and Watt [13].

4.7   Methods for the Solution
of Boundary-Value Problems

As mentioned previously, any ordinary differential equation that has constraints imposed at values of the independent variable other than at zero is called a boundary-value problem. If only one condition were specified in a boundary-value problem (as would be the case for a first-order differential equation), the independent variable could be adjusted by a change of variable to transform the boundary-value problem into an initial-value problem. For this reason it only makes sense to deal with boundary-value problems that are of second order or higher.

For simplicity in this discussion, we speak in terms of a second order equation:

$$\frac{d^2y}{dx^2} = f(x, y, y')$$

with boundary conditions:

$$y(a) = A \qquad \text{and} \qquad y(b) = B,$$

even though equations of higher order can be treated by the same techniques. Solution methods for this system fall into two general categories:

1. those techniques that rely on reducing the problem to that of solving multiple initial-value problems; and

2. those techniques that employ a finite difference form of the differential equation.

Initial-Value Methods (Shooting Methods)

If the second-order differential equation is linear of the form:

$$y'' = f_1(x)y' + f_2(x)y + f_3(x)$$

$$y(a) = A, \qquad y(b) = B$$

it may be reduced to an initial-value problem by means of the initial conditions:

$$y(a) = A \qquad \text{and} \qquad y'(a) = \alpha_1.$$

Once a solution $y_1(x)$ is known, a different set of boundary conditions:

$$y(a) = A \qquad \text{and} \qquad y'(a) = \alpha_2$$

may be applied to achieve a second solution $y_2(x)$. If $y_1(b) = \beta_1$ and $y_2(b) = \beta_2$, then if $\beta_1 \neq \beta_2$, the solution:

$$y(x) = \frac{1}{\beta_1 - \beta_2} [(B - \beta_2)y_1(x) + (\beta_1 - B)y_2(x)]$$

satisfies both of the original boundary conditions.

If the differential equation is nonlinear, a series of initial-value problems may be solved using successive values of α in the initial conditions:

$$y(a) = A \qquad \text{and} \qquad y'(a) = \alpha$$

in an attempt to find a solution that will ultimately satisfy y(b) = B. In
such a process, interpolation can often be used to suggest an orderly
progression for α that will minimize wasted effort. Unfortunately, this
technique is inherently inefficient and is not recommended as a substitute
for some of the advanced methods available in the literature.

Finite Difference Methods

The advantage to using a finite difference approach is that the
solution of the boundary-value problem may be reduced to that of solving
a system of algebraic equations. In order to treat the two-point
boundary-value problem

$$y" = f(x, y, y')$$

$$y(a) = A, \qquad y(b) = B,$$

the interval from a to b may be divided into "n" equal parts:

$$x_i = x_0 + ih \qquad i = 1, 2,..,n$$

where x_0 = a, x_n = b and

$$h = \frac{b - a}{n}.$$

The x_i points are called "node" points and represent locations at which the
ordinate values y_i of the solution are desired. Using these node locations
and finite difference relationships for the derivatives:

$$y'(x_i) = \frac{1}{2h} (y_{i+1} - y_{i-1})$$

$$y"(x_i) = \frac{1}{h^2} (y_{i+1} - 2y_i + y_{i-1})$$

the differential equation can be written as a "difference equation." (It
should be noted that various types of finite difference forms exist for the
expression of derivatives. The topic of difference relationships is
discussed in more depth in the next chapter.) If this difference equation
is written for i = 1, 2,..,n and the two boundary conditions are used, the
problem is reduced to a system of n - 1 equations in n - 1 algebraic
unknowns y_i. If the original differential equation is linear, the problem
will give rise to the simultaneous solution of a set of linear algebraic
equations. If, on the other hand, the original differential equation is
nonlinear, the problem solution will become that of solving a system of
simultaneous, nonlinear algebraic equations. Although the solution of
algebraic equations is well understood, the solution of boundary - value
problems by the method of finite differences is difficult to systematize into

89

a "prepackaged" computer subroutine, because the formulation of each problem depends on the nature of the specific differential equation being considered.

Example 4-3

Suppose that it is desired to solve the differential equation:

$$y'' = 2x + 3y$$

subject to:

$$y(o) = 0, \qquad y(1) = 1$$

using h = 0.2. The differential equation can be written in finite difference form as:

$$\frac{1}{0.04} (y_{i+1} - 2y_i + y_{i-1}) = 2x_i + 3y_i.$$

This formula and the boundary conditions can be used to write the following system of four linear equations in four unknowns:

$$-2.12y_1 + y_2 \qquad = 0.016$$

$$y_3 - 2.12y_2 + y_1 = 0.032$$

$$y_4 - 2.12y_3 + y_2 = -0.064$$

$$-2.12y_4 + y_3 \qquad = -0.936$$

The following table below presents a comparison of the results of solving this system and the exact solution, which is:

$$y(x) = \frac{5}{3} \frac{\sinh\sqrt{3x}}{\sinh\sqrt{3}} - \frac{2}{3}.$$

x	$y_{numerical}$ solution	y_{exact}
0.0	0.0	0.0
0.2	0.0827	0.0818
0.4	0.1912	0.1897
0.6	0.3548	0.3529
0.8	0.6088	0.6073
1.0	1.0000	1.0000

4.8 Considerations in the Selection of an Algorithm for Solving Ordinary Differential Equations

Although it is impossible to state universal rules that will guide the user in selecting a best method for solving a given ordinary differential equation, a few basic guidelines do exist:

1. Consider the Type of Problem. If the problem to be solved is an initial-value problem, many different types of prepackaged, computer subroutines are available to implement the solution. If the problem is a boundary-value problem, the user may have to write his or her own software.

2. Consider the Complexity of the Differential Equation. If an initial-value problem is quite complicated and the evaluation of $f(x,y)$ is rather involved, a predictor-corrector method is usually preferred because it requires only two evaluations of $f(x,y)$ for each step rather than the four required for Runge-Kutta methods. For multipurpose uses where the evaluation of $f(x,y)$ is not difficult, the Runge-Kutta method is often more convenient.

3. Consider the Time Involved in Solving the Problem. If computer time cost is a premium factor, the best method to use is a predictor-corrector method. If on the other hand, the user preparation time is the critical factor, a method such as the Runge-Kutta method may be preferred.

4. Consider the Required Accuracy. In general, the higher order a method is, the higher will be the resulting accuracy. This principle is true up to a point. Finite difference approximations for higher-order derivatives tend to behave increasingly badly. Because of this fact, the error available in a fifth-order method is about the same as available through more tedious, higher-order techniques. Since there usually exists a tradeoff between computational efficiency and computational accuracy, the user should exercise care in selecting both the order of the method and the step size. For this reason algorithms that allow automatic adjustment of the step size or order of the method are well worth using.

5. Consider Past Performance. Previous success, or lack of success, when applying a particular algorithm to a given type of engineering problem can often provide valuable insight into the selection of suitable software. This is especially true when the differential equation system is stiff, because the user will usually not know that a system is stiff until after he or she has experienced poor performance of a given algorithm.

4.9 Available Packages for Modeling Engineering Systems

Almost every computer software package in existence contains one or more more subroutines to facilitate the solution of ordinary differential equations of the initial-value type. To list all of these would certainly become counterproductive since either a Runge-Kutta method or a predictor-corrector method is used in nearly every case.

Because the problem of simulation of engineering systems tends to occur quite frequently in engineering problem solving, a number of special-purpose simulation programs have been developed. These simulation programs are classified as either continuous or discrete depending on whether they are applied to problems that are continuous in nature or to problems that are defined only for discrete values. Examples of continuous system applications would be vibration analysis, circuit analysis, etc. Examples of discrete system applications would be inventory systems, traffic flow systems, customer behavior, etc. Because discrete systems tend to involve variables that change randomly, they are said to be stochastic. In contrast, continous systems tend to be described in terms of precise inputs and precise outputs and are referred to as deterministic. Table 4-1 summarizes the characteristics of a few well known simulation packages.

Table 4-1   Simulation Languages for Digital Computers

Name	Continuous/ Discrete	Originating Organization	Source Reference
CSMP	C	IBM Corporation	[29]
CSSL	C	Control Data Corporation	[30]
DYSAC	C	University of Wisconsin	[15]
GASP	C-D	U.S. Steel Corporation	[24]
GPSS	D	IBM Corporation	[11]
MIDAS	C	Wright-Patterson Air Force Base	[23]
MIMIC	C	Wright-Patterson Air Force Base	[28]
PACTOLUS	C	IBM Corporation	[2]
SL-1	C	Xerox	[33]

Early versions of continuous system modeling programs were special packages that enabled the digital computer to be programmed and operated in block diagram form as if it were an analog computer. For this reason these programs were often referred to as digital-analog simulators. Later versions of these simulation programs have been more general and tend to be more like special purpose programming languages that allow greater versatility in the setup and operation of the simulation process. The integration process in continuous system modeling programs is accomplished by Runge-Kutta or predictor-corrector methods. Interest in the development and use of simulation programs has grown at a rapid rate in the past few years, and a number of excellent simulation packages are now available to the engineering user. These packages have the time-saving advantage that they can easily be used to model complex engineering problems that would otherwise require considerable programming effort. Although it is beyond the scope of this text to describe the use of simulation programs, the interested reader should consult some of the references listed at the end of this chapter.

Problems

4.1 The mass shown in the following figure is permitted to swing on a slender uniform rod. Prepare and run a computer program that will simulate the motion of this device for one complete swing cycle if:

$$\ddot{\theta} + (g/L) \sin\theta = 0$$

$$\theta(0) = \pi/4$$

$$g = 9.8 \text{ m/s}^2$$

$$\dot{\theta}(0) = 0$$

$$L = 0.10 \text{ m}$$

In cases where the angular deflection is small, this differential equation is sometimes linearized by setting $\sin\theta = \theta$. Compare your results with this linear case.

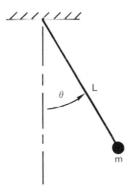

4.2 Solve Example 4-2 by a predictor-corrector method and compare your results with those found by the Runge-Kutta method.

4.3 The mass shown in the following figure moves on a flat surface with dry friction damping. If the mass is 4.5 kg, the spring constant is k = 175 N/m, and the coefficient of friction is 0.3; find and plot the resulting motion for $0 \leq t \leq 2$ sec. using the initial conditions:

$$x(0) = 7.5 \text{ cm}$$

$$\dot{x}(0) = 0.0 \text{ cm/sec.}$$

Use: (a) a Runge-Kutta Method and
      (b) a predictor-corrector method.

Which method uses less computer time? Why?

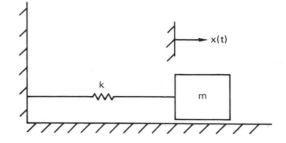

4.4 The linkage between the key and striker hammer for the piano has been studied with a view toward understanding how its operation may be improved. Oledzki[†] has proposed the following model that is nonlinear:

$$\ddot{x}_1 = \frac{F}{m_1} - (x_1 - x_2)(k_1 + k_2 x_2) \; \frac{1}{m_1} \; - \frac{k_3}{m_1}(x_1 - x_2)^2$$

$$\ddot{x}_2 = \frac{b\dot{x}_2{}^2 + (x_1 - x_2)(k_1 + k_2 x_2) - 0.5(x_1 - x_2)^2 k_3}{a - b x_2}$$

In these differential equations the parameter x_1 describes the downward displacement of the piano key, the parameter x_2 describes the forward motion of the striker hammer, and the parameter F represents the downward force applied to the key.

The numerical values of the coefficients for an upright piano are estimated to be:

$m_1 = 0.074$ kg

$a_0 = 0.406$ kg

$b_0 = 18.3$ kg m^{-1}

$k_{1,0} = 1.16 \times 10^4$N m^{-1}

$k_{2,0} = 0.525 \times 10^6$N m^{-2}

$k_{3,0} = 1.1 \times 10^6$N m^{-2}

$0 < F < 80$ N.

Prepare and run a computer program that will simulate the motion of this system if:

[†]Oledzki, A., "Dynamics of Piano Mechanisms," Mech. and Mach. Theory 7, (Winter 1972): 373-85.

$$F = 80 \text{ N}$$

$$0 \leq t \leq 30 \text{ msec.}$$

4.5 A home heating system can be described in terms of the following differential equation:

$$Q_{in} - Q_{out} = 250 \frac{dT_c}{dt},$$

where: Q_{in} = input heat from the furnace (J/sec)

Q_{out} = heat loss to environment (J/sec)

T_c = house inside temperature (°C)

t = time (sec).

This type of system is often equipped with an "on-off" control thermostat that turns the furnace on or off depending on the difference between the desired temperature T_d and the actual temperature T_c.

If the heat loss on a day when the outside temperature is 0°C is:

$$Q_{out} = 500(T_c - 0°C),$$

and if the controller performs as follows:

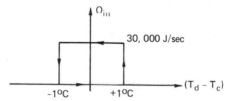

write a computer program that will simulate this system for T_d = 22°C assuming that the system is initially at temperature T_c = 10°C. What will be the frequency of the limit cycle for this system?

4.6 Solve the following boundary-value problem:

$$y'' = 2x^2 + 3y^3 + 4xy$$

$$y(0) = 0.$$

$$y(1) = 1.0$$

using h = 0.2 and then using h = 0.1. Do you get different results? If so, why?

4.7 For the swinging pendulum of Problem 4.1 it is desired to reach θ = $\pi/4$ at t = 1.0 sec. If $\dot{\theta}(0)$ = 0, what starting angle should be used to achieve this result?

4.8 The differential equation for the deflection of the constant cross section beam is:

$$\text{curvature} = \frac{M(x)}{E I} = \frac{PL^2}{EI} \left(\frac{1}{L} - \frac{x}{L^2} \right)$$

If the initial conditions are:

$$y(0) = 0$$

$$y'(0) = 0$$

and if the length and load parameters are:

$$L = 1.0 \text{ m}$$

$$PL^2/EI = 2.0,$$

write a computer program that will generate the elastic curve $y(x)$ for this beam using:

a)   the exact expression for curvature:

$$\text{curvature} = \frac{y''}{[1 + y'^2]^{3/2}}$$

(b) the approximation:

$$\text{curvature} = y''$$

Compare the results.   How adequate is the approximation?

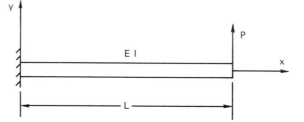

4.9   For the piano mechanism described in Problem 4.4, it is desired to have the value of x_1 at t = 45 m sec equal to 8 mm.   What striking force F will be required to achieve this?

4.10 If the damper in Example 4-2 were removed, find and plot the relationship between free vibration frequency w_f and initial displacement x_0 for:

$$0 \leq x_0 \leq 30 \text{ cm}.$$

(Use $x(0) = x_0$ and $\dot{x}(0) = 0.0$ cm/sec.)

References

1.   Bennett, A. W. Introduction to Computer Simulation.   New York: West Publ. Co., 1974.

2.   Brennan, R. D., and Sano, H.   "Pactolus: A Digital Analog Simulator Program for IBM 1620."   Proc. Fall Joint Computer Conference (1964): 299-312.

3.   Ceshino, F., and Kuntzmann, J. Numerical Solution of Initial-Value Problems.   Englewood Cliffs, N.J.:   Prentice-Hall, Inc., 1966.

4.   Dahlquist, G., and Bjorck, A. Numerical Methods. Englewood
      Cliffs, N.J.: Prentice-Hall, Inc., 1974.

5.   Dorn, W. S. and McCracken, D. D. Numerical Methods with
      FORTRAN IV Case Studies. New York: John Wiley & Sons, 1972.

6.   Forsythe, G. E., Malcolm, M. A., and Moler, C. B. Computer
      Methods for Mathematical Computations. Englewood Cliffs, N.J.:
      Prentice-Hall, Inc., 1977.

7.   Gear, C. W. Numerical Initial Value Problems in Ordinary Differential
      Equations. Englewood Cliffs, N.J.: Prentice-Hall, Inc., 1971.

8.   Gear, C. W. "Hybrid Methods for Initial-Value Problems in Ordinary
      Differential Equations." J. SIAM Numer. Anal., (Ser. B), 2, (1964):
      69-86.

9.   Gear, C. W., and Watanabe, D. S. "Stability and Convergence of
      Variable Order Multistep Methods." SIAM J. Numer. Anal., 11 (Oct.
      1974): 1044-58.

10.   Gear, C. W., and Tu, K. W. "The Effect of Variable Mesh Size on
      the Stability of Multistep Methods." SIAM J. Numer. Anal. 11 (Oct.
      1974): 1025-1043.

11.   Greenberg, S. GPSS-Primer. New York: Wiley Interscience, 1972.

12.   Grove, W. E. Brief Numerical Methods. Englewood Cliffs, N.J.:
      Prentice-Hall, Inc., 1966.

13.   Hall, G., and Watt, J. M. Modern Numerical Methods for Ordinary
      Differential Equations. Oxford, England: Clarendon Press, 1976.

14.   Hamming, R. W. "Stable Predictor-Corrector Methods for Ordinary
      Differential Equations." J. Assoc. Comput. Mach. 6 (1959): 37-47,
      1959.

15.   Hurley, J. R. and Skiles, J. J. "DYSAC, A Digitally Simulated
      Analog Computer" Proc. Spring Joint Computer Conf., (1963): 69-82.

16.   IBM System/360 Scientific Subroutine Package (SSP). 360A-CM-03X
      Version 3, 6th Ed., March 1970.

17.   Jennings, W. First Course in Numerical Methods. New York:
      MacMillan Co., 1964.

18.   Kochenburger, R. J. Computer Simulation of Dynamic Systems.
      Englewood Cliffs, N.J.: Prentice-Hall, Inc., 1972.

19.   Lapidus, L., and Seinfeld, J. H. Numerical Solution of Ordinary
      Differential Equations. New York: Academic Press, 1971.

20.   Markowitz, H. M., Hausner, B., and Kerr, H. W. SIMSCRIPT: A
      Simulation Programming Language. Englewood Cliffs, N.J.: Prentice-
      Hall, Inc., 1963.

21.   McCalla, T. R. Introduction to Numerical Methods and FORTRAN
      Programming. New York: John Wiley & Sons, 1967.

22.   Pall, G. A. Introduction to Scientific Computing. New York:
      Appleton-Century-Crofts Educational Division, Meredith Corporation,
      1971.

23. Peterson, H. E., Sansom, F. J., Harnett, R. T., and Warshawsky, L. M. "MIDAS: How it Works and How It's Worked." Proc. Fall Joint Computer Conf., (1964): 313-24.

24. Pritsker, A. B., and Hurst, N. R. "GASP IV: A Combined Continuous-Discrete FORTRAN Based Simulation Language." Simulation, 21 (1973): 67-75.

25. Ralston, A. A First Course in Numerical Analysis. New York: McGraw-Hill Book Co., 1964.

26. Ralston, A. and Wilf, H. S. Mathematical Methods for Digital Computers. New York: John Wiley & Sons, 1967.

27. Richardson, L. F. and Gaunt, J. A. "The Deferred Approach to the Limit." Trans. Roy. Soc. London, 226A (1927): 300.

28. Sansom, F. J. MIMIC Programming Manual, Tech. Rept. SEG-TR-67-31, Wright-Patterson Air Force Base, Ohio, 1967.

29. Speckhart, F. H., and Green, W. L. A Guide to Using CSMP-the Continuous System Modeling Program. Englewood Cliffs, N.J.: Prentice-Hall, Inc., 1976.

30. The SCI Simulation Software Committee, "The SCI Continuous System Simulation Language (CSSL)." Simulation (Dec. 1967): 281-303.

31. Tiechroew, D., Lubin, F. W., and Truitt, T. D. "Discussion of Computer Simulation Techniques and Comparison of Languages." Simulation, 9 (1967): 181-190.

32. Williams, P. W. Numerical Computation. New York: Barnes & Noble, 1972.

33. Xerox SL-1 Simulation Language, Reference Manual 90-16-76 B, Xerox Data Systems, Feb. 1972.

5 COMPUTER SOLUTION OF PARTIAL DIFFERENTIAL EQUATIONS

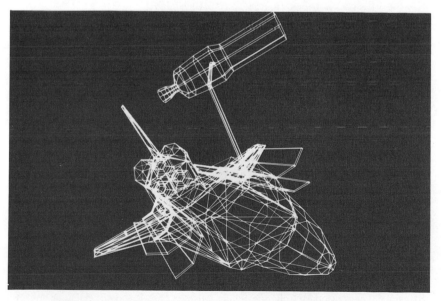

Finite element representation of the space shuttle. [Computer graphics output courtesy of Evans & Sutherland.]

5.1　Introduction

For a large variety of engineering problems, the dependent variable is expressed in terms of several independent variables. Such problems inherently give rise to the need for partial derivatives in the description of their behavior. The study of the differential equations arising from these problems constitutes the field of "partial differential equations."

It is an unfortunate fact that a large number of important partial differential equation problems in engineering cannot be solved in exact form. Numerical methods are often used to overcome this difficulty. Although many good alternative methods exist for the solution of ordinary differential equations, the variety of alternatives available for solution of partial differential equations is limited to either finite difference methods or finite element methods. In this chapter the topic of numerical solution of partial differential equations is discussed in terms of the engineering applications of these two methods. The common types of partial differential equations are classified, and the rationale for their numerical solution is presented.

5.2　Categories of Partial Differential Equations

Partial differential equations are frequently classified in terms of their mathematical form (such as elliptic, parabolic, etc.) or in terms of the type of problem to which they apply (for example, the diffusion equation, the wave equation, etc.). In order to extract solution techniques from both the theoretical literature of mathematics and the applications literature of technology, the engineer must be familiar with both classification schemes.

In the mathematical sense, the second-order partial differential equation in two independent variables:

$$A(x,y)\frac{\partial^2 f}{\partial x^2} + B(x,y)\frac{\partial^2 f}{\partial x \partial y} + C(x,y)\frac{\partial^2 f}{\partial y^2} + E(x,y,f,\frac{\partial f}{\partial x},\frac{\partial f}{\partial y}) = 0$$

can be classified according to the nature of the functions A, B, and C. These functions depend on the independent variables x and y. If $B^2 - 4AC < 0$, the equation is elliptic. If $B^2 - 4AC = 0$, the equation is parabolic. If $B^2 - 4AC > 0$, the equation is hyperbolic. This classification scheme is rather interesting because the values of A, B, and C depend on the independent variables. Thus it is possible for a partial differential equation to change class within the different regions of the domain for which the problem is defined.

In terms of boundary conditions, the second-order partial differential equation can have constraints in the form of boundary values, initial values, or combinations of these. Problems in the elliptic class are equilibrium problems and are described in terms of a closed region having boundary conditions prescribed at every point on the region boundary. Problems in the parabolic and hyperbolic class are "propagation" problems and can have prescribed boundary conditions on some boundaries, can have initial conditions along some boundaries, and can also have open-ended regions into which the solution propagates.

In the engineering sense, partial differential equations frequently occur in a few common forms. A list of the more familiar of these is presented in Table 5-1. In this table the "∇^2" operator is known as the Laplace operator. In terms of one independent variable:

$$\nabla^2 f = \frac{\partial^2 f}{\partial x^2}$$

In terms of two independent variables:

$$\nabla^2 f = \frac{\partial^2 f}{\partial x^2} + \frac{\partial^2 f}{\partial y^2}$$

The "∇^4" operator is known as the biharmonic operator and is expressed in two independent variables as:

$$\nabla^4 f = \frac{\partial^4 f}{\partial x^4} + 2\,\frac{\partial^4 f}{\partial x^2 \partial y^2} + \frac{\partial^4 f}{\partial y^4}$$

5.3 Numerical Solution of Partial Differential Equation Problems by the Finite Difference Method

The finite difference method for the solution of partial differential equation problems is based on the use of finite difference approximations for derivatives. The procedure followed is in many ways similar to that presented in the previous chapter for the solution of ordinary boundary value problems. It consists of the three steps illustrated in the diagram of Figure 5-1. The steps are implemented as follows. First, the solution domain is divided into a grid of "node" points. This grid is uniformly spaced, and its shape reflects the nature of the problem and its boundary conditions. Next, the governing partial differential equation is written in terms of the most convenient coordinate system available and is transformed into a partial difference equation by means of finite difference approximations to the derivatives involved. This difference formula will be

Table 5-1   Common Types of Partial Differential Equations in Engineering

Equation Type	Equation Form	Example Application
Laplace's Equation:	$\nabla^2 f = 0$	Steady-state flow of heat & fluids.
Poisson's Equation:	$\nabla^2 f = -k$	Heat transfer with internal heating.
The Diffusion Equation:	$\nabla^2 f = \dfrac{1}{h^2}\dfrac{\partial f}{\partial t}$	Nonequilibrium states of heat conduction.
The Wave Equation:	$\nabla^2 f = \dfrac{1}{c^2}\dfrac{\partial^2 f}{\partial t^2}$	Propagation of acoustic waves.
The Biharmonic Equation:	$\nabla^4 f = F(x,y)$	Deformation of plates.

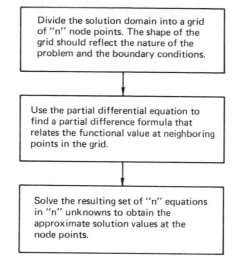

Divide the solution domain into a grid of "n" node points. The shape of the grid should reflect the nature of the problem and the boundary conditions.

Use the partial differential equation to find a partial difference formula that relates the functional value at neighboring points in the grid.

Solve the resulting set of "n" equations in "n" unknowns to obtain the approximate solution values at the node points.

Figure 5-1   Methodology for the numerical approximation of partial differential equation problems by the finite difference method.

used to describe the functional relationship between nearby points in the grid. The difference equation is written for every point in the grid, and the result is a set of "n" equations in "n" unknowns. Finally, the system of "n" equations and "n" unknowns is solved using a numerical technique. Although this three-phase process may seem simple and straightforward, considerable variation in grid types, grid sizes, partial differential equations, finite difference approximations to these equations, and solution techniques for the resulting equation system makes the topic of computer solution of partial differential equations an extremely diverse and interesting study. Details of the alternatives for each of the three phases of the solution process are now discussed.

5.4   Grid Patterns for Use in Approximating Partial Differential Equations

Although the partial differential equations presented thus far have all been expressed in Cartesian coordinate form, it is true that many engineering problems are best expressed in terms of other coordinate systems that possess special geometric features duplicating the physical characteristics of the problem being considered. Examples of common engineering coordinate systems include Cartesian, cylindric, and spherical coordinates. Moon and Spencer [8] present a listing of 40 unique coordinate systems that can be used to describe and quantify engineering problems.

The most commonly used grid systems for partial differential equation formulation are presented in Figure 5-2. These are the rectangular,

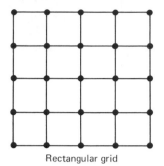

Rectangular grid

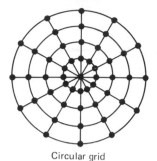

Circular grid

Triangular grid

Skew grid

Figure 5-2 Commonly used grid systems for partial differential equations.

circular, triangular, and skew grids. Within this family the rectangular grid is most frequently used. Quite often irregularly shaped boundaries such as those represented in Figure 5-3 are encountered. Even though these boundaries cannot be exactly represented in terms of any single grid pattern, special techniques are available to allow modification of the standard grids to accommodate the boundaries. For problems of the "propagation" type, the grid may be visualized as a pattern that extends into an open domain as far as is necessary to extract the desired information from a given problem situation.

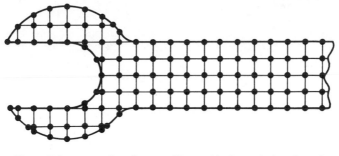

Figure 5-3   Example of a problem with irregularly shaped boundaries.

5.5   Finite Difference Representations for Partial Derivatives

A Taylor series expansion of a function $f(x,y)$ of two independent variables can be used to approximate the partial derivative

$$\frac{\partial f}{\partial x} = \frac{[f(x_i + h,y_j) - f(x_i - h,y_j)]}{2h}.$$

In this expression "h" represents a small change in the "x" value from the "i" location. In this expression the terms that have been neglected are of the order of h^2, and the expression is called a central difference formula since it is symmetrically placed about the point of interest (x_i,y_j). It is convenient to think of this finite difference representation as referring to three adjacent points in a two-dimensional grid of spacing "h" as shown in Figure 5-4. In this figure the "j" subscript applies to the "y" independent variable, and the "i" subscript applies to the "x" variable. For convenience the notation $f(x_i + h,y_j)$ can be simplified to $f_{i+1'j}$. Using this same notation and the Taylor series expansion, finite difference representations for other partial derivatives can be found. Examples of the more common of these are:

$$\frac{\partial f}{\partial x} = \frac{f_{i+1'j} - f_{i-1'j}}{2h}$$

$$\frac{\partial f}{\partial y} = \frac{f_{i,j+1} - f_{i,j-1}}{2h}$$

105

$$\frac{\partial^2 f}{\partial x^2} = \frac{f_{i+1,j} - 2f_{i,j} + f_{i-1,j}}{h^2}$$

$$\frac{\partial^2 f}{\partial y^2} = \frac{f_{i,j+1} - 2f_{i,j} + f_{i,j-1}}{h^2}$$

$$\frac{\partial^2 f}{\partial x \partial y} = \frac{f_{i+1,j+1} - f_{i-1,j+1} - f_{i+1,j-1} + f_{i-1,j-1}}{4h^2}$$

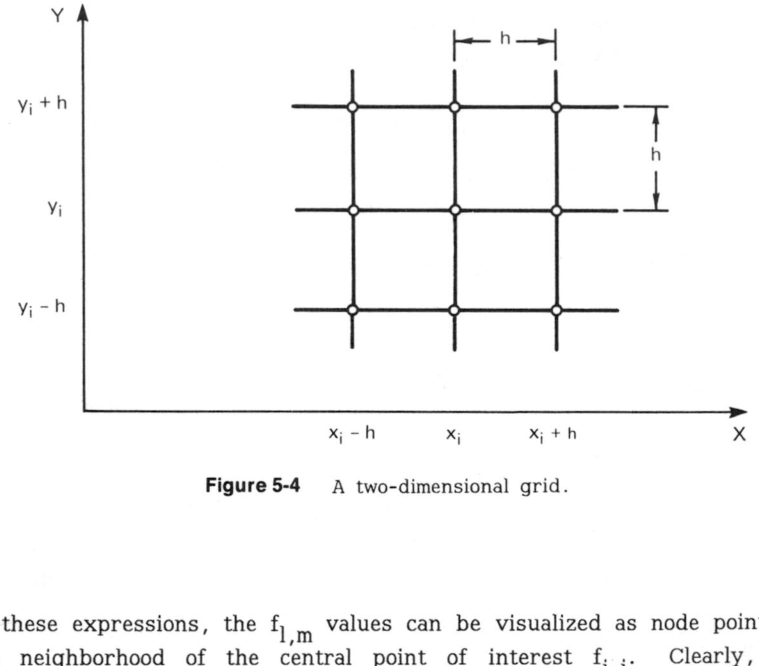

Figure 5-4   A two-dimensional grid.

In these expressions, the $f_{l,m}$ values can be visualized as node points in the neighborhood of the central point of interest $f_{i,j}$. Clearly, the significant features contained in these finite difference representations are the direction of interest (either x or y) and the coefficients of the various $f_{l,m}$ terms. A streamlined way to represent this information is through use of a "computational template," which is a diagram that shows how the various neighborhood points in the grid contribute to the derivative of interest. A listing of the computational templates for various types of common derivatives is presented in Figure 5-5. These computational templates become the basic building blocks for the construction of more complex templates to be used for the differential equations. The addition process for derivatives is achieved by the superposition of the necessary

106

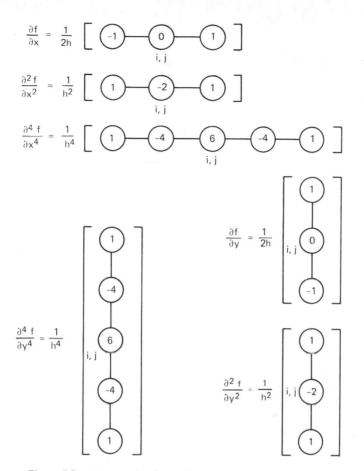

Figure 5-5   Computational templates for common derivatives.

computational templates. Using this technique, the computational templates for $\nabla^2 f$ and $\nabla^4 f$ can be assembled and are presented in Figure 5-6. In each case these templates have errors on the order of "h^2". It is well worth mentioning that more accurate templates (that have smaller errors) can be found if the user is willing to utilize additional node points in the computation process. Each of the templates constructed so far has been based on central difference relationships. Sometimes a forward- or backward-difference expression is used in order to minimize the propagation of error. Indeed, the computational template concept must be applied with care since it is possible to formulate a partial differential equation approximation in an unstable form. In this sense, an unstable system is one in which an error from any source does not diminish as time increases. Propagation problems are particularly susceptible to stability difficulties.

107

$h^2\nabla^2 =$

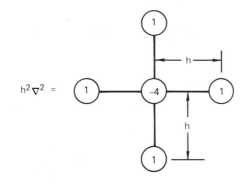

$h^4\nabla^4 =$

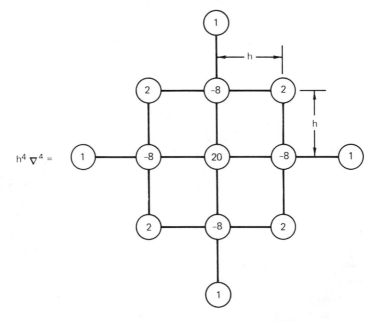

Figure 5-6 Computational templates.

Irregular Boundaries

As mentioned previously, the computational template for a given partial differential equation can be modified to accommodate irregular boundaries by recognizing the contribution of the boundary point to the central difference form of the derivatives. As an example consider the Laplacian template with irregular boundaries, as illustrated in Figure 5-7(a). For this situation $a < 1$ and $b < 1$. The second partial derivatives may be formed in terms of the boundary points as:

$$\frac{\partial^2 f}{\partial x^2} = \frac{\dfrac{f_a - f_{i,j}}{ah} - \dfrac{f_{i,j} - f_{i-1,j}}{h}}{\frac{1}{2}(ah + h)}$$

$$\frac{\partial^2 f}{\partial y^2} = \frac{\dfrac{f_b - f_{i,j}}{bh} - \dfrac{f_{i,j} - f_{i,j-1}}{h}}{\frac{1}{2}(bh + h)} \ .$$

These two derivatives may be added together to get $\nabla^2 f$. The result, after a bit of manipulation is:

$$\nabla^2 f = \frac{2}{h^2}\left[\frac{f_{i-1,j}}{1 + a} + \frac{f_a}{a(1 + a)} + \frac{f_b}{b(1 + b)} + \frac{f_{i,j-1}}{1 + b} - \frac{(a + b)}{ab} f_{i,j}\right]$$

The resulting computational template is presented in Figure 5-7(b). This expression reduces to the traditional Laplacian template if $a = b = 1$. Although the procedure demonstrated here was applied to the Laplacian operator, the method is applicable to any partial differential equation that can be represented in template form.

5.6   Iterative Solution Techniques

Once the "computational template" is applied to each of the "n" node points in a grid, the result is a system of "n" equations. The system of equations may be linear depending on the nature of the differential equation being used. In the linear case it is necessary to solve a system of equations of the form:

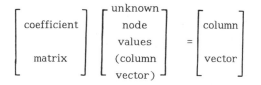

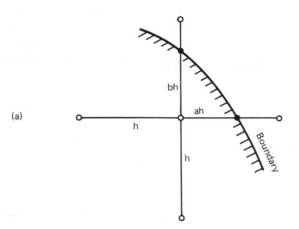

(a)

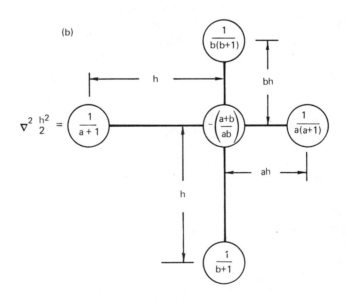

(b)

$$\nabla^2 \frac{h^2}{2} =$$

Figure 5-7 Irregular boundaries.

110

In most cases the coefficient matrix will be sparsely populated (it will have many zero terms) because most computational stencils use only a few adjacent points rather than every point in the grid. In general the techniques used to solve the system are classified as either direct or iterative. The direct methods find the exact solution of the system in a finite number of steps. An example of direct method might be Cramer's rule for the solution of simultaneous linear algebraic systems. As a general rule, for larger systems of equations, the direct methods are not efficient in effort or in computer storage. For this reason the iterative methods are more often used.

Iterative methods involve the repeated application of a simple procedure or algorithm, and they provide answers that gradually approach the true solution. The iterative procedure begins with an initial approximation to the solution and then successively modifies the node values until they reach the solution to the desired accuracy. The speed of convergence for a given method depends strongly on the degree of accuracy of the initial approximation. Thus intuitive engineering judgment can greatly influence the efficiency of the computational process.

Iterative methods are generally classified into two broad categories known as point iterative methods and block iterative methods. In point iterative methods, an algorithm is used to modify the approximate solution at a single point in the grid domain. In block methods a whole group of node values is modified simultaneously. Some examples of common point iterative methods are now presented.

Method of Simultaneous Displacements

The method of simultaneous displacements [1], also known as iteration by total steps or the Jacobi method, is the simplest of all iterative techniques. In spite of its simplicity, it is seldom used because it is quite slow to converge. Nevertheless, it is discussed here because it provides a basis for understanding other methods. For the method of simultaneous displacements, the procedure is as follows. A completely new grid of values is assembled, node by node, using the computational template and the old values. Once the new grid of values is complete, it is used to replace the value of every point in the grid all at once. Since the grid is not modified until new values for all points have been computed, the order of the calculation process is unimportant. The procedure is said to be complete when the change in value at every node in the grid is smaller than some predetermined value. In order to apply any iterative technique, it is first necessary to start with initial approximations. These may be obtained by any reasonable method. Frequently, linear interpolation is employed to make initial estimates of the node values. Naturally, the closer the initial approximations are to the solution, the fewer will be the number of iterations required to complete the iterative process. In order to illustrate the method of simultaneous displacements, it is useful to consider an example problem.

Example 5-1

Suppose that is is desired to find the two-dimensional steady-state temperature distribution in a square plate subject to the boundary conditions:

$$\text{at} \quad x = 0.0 \quad T = 0.$$

$$x = 1.0 \quad T = 100.$$

$$y = 0.0 \quad T = 100 \ x \quad \text{and}$$

$$y = 1.0 \quad T = 100 \ x^2.$$

This problem will be defined by the two-dimensional Laplacian equation:

$$\nabla^2 T = \frac{\partial^2 T}{\partial x^2} + \frac{\partial^2 T}{\partial y^2} = 0$$

In order to set up this problem, a two-dimensional grid of spacing h = 0.25 is drawn on the plate. This grid gives rise to 25 node points, 16 of which have temperature values that are known by virtue of the boundary conditions. The grid system is illustrated in the following figure. The unknown node points

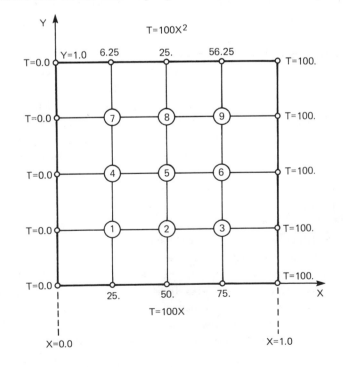

112

are identified by numbers on the diagram. The problem now becomes that of finding the appropriate values at each of the nine internal node points.

The new grid temperature value at a point $T_{i,j}$ may be found from the template:

$$T_{i,j} = 0.25(T_{i-1,j} + T_{i+1,j} + T_{i,j-1} + T_{i,j+1})$$

Linear interpolation is used to establish initial values for the temperatures in the grid and the iterative method is started. The results found after 30 iterations are presented in the table on the next page. Since this example problem has only a few node points, it is possible to find the actual solution by a direct method. The direct solution is presented in Table 5-2 for comparison. Once a satisfactory solution is found, linear interpolation may be used to find the temperature values at locations between node points.

Method of Successive Displacements

One of the characteristics of the method of simultaneous displacements that makes it converge so slowly is the fact that the impact of an improvement is not exploited until every point in the grid has been recalculated (i.e., when the replacement is made). The method of successive displacements [1] also known as the method of iteration by single steps or the Liebman procedure makes use of an improvement as soon as it is available. Thus an improvement in the value at node "1" will be used in the calculation for a new node value at point "2" and so on. Clearly, for this method, the order in which the node points are used will influence the way that the problem proceeds. Since this method uses new information as soon as it is available, it tends to converge more rapidly. The results achieved through the application of this method to Example 5-1 are now presented in Table 5-3. The output of this example clearly indicates that the rate of convergence for this method is better than the previous method. Fewer iterations are required to achieve the same degree of accuracy.

Method of Successive Over-Relaxation

Relaxation methods can be used to solve linear algebraic equation systems. The basis for these methods is a technique that successively reduces the "residual" at each of the node points. (The residual is the amount by which a node value differs from the correct solution as predicted by the computational template.) The pioneer researcher into relaxation methods for partial differential equation solution was Southwell [14]. In his efforts Southwell discovered that it is often useful to change a given node value by more than would be necessary to reduce a given residual to zero. This "over-relaxation" method is based on a linear

113

Table 5-2 The Results of Example 5-1 by the Method of Simultaneous Displacements

Iteration number	T_1	T_2	T_3	T_4	T_5	T_6	T_7	T_8	T_9
Start	20.313	43.750	70.313	15.625	37.500	65.625	10.983	31.250	60.938
1	21.094	44.531	71.094	17.188	39.063	67.188	13.281	33.594	63.281
2	21.680	45.313	71.680	18.359	40.625	68.359	14.258	35.156	64.258
3	22.168	45.996	72.168	19.141	41.797	69.141	14.941	36.035	64.941
4	22.534	46.533	72.534	19.727	42.578	69.727	15.356	36.670	65.356
5	22.815	46.912	72.815	20.117	43.164	70.117	15.662	37.073	65.662
6	23.007	47.198	73.007	20.410	43.555	70.410	15.860	37.372	65.860
7	23.152	47.392	73.152	20.605	43.848	70.605	16.008	37.569	66.008
8	23.249	47.538	73.249	20.752	44.043	70.752	16.106	37.716	66.106
9	23.322	47.635	73.322	20.850	44.189	70.850	16.179	37.814	66.179
10	23.371	47.709	73.371	20.923	44.287	70.923	16.228	37.887	66.228
11	23.408	47.757	73.408	20.972	44.360	70.972	16.265	37.936	66.265
12	23.432	47.794	73.432	21.008	44.409	71.008	16.289	37.973	66.289
13	23.451	47.818	73.451	21.033	44.446	71.033	16.308	37.997	66.308
14	23.463	47.837	73.463	21.051	44.470	71.051	16.320	38.015	66.320
15	23.472	47.849	73.472	21.063	44.488	71.063	16.329	38.027	66.329
16	23.478	47.858	73.478	21.072	44.501	71.072	16.335	38.037	66.335
17	23.483	47.864	73.483	21.078	44.510	71.078	16.340	38.043	66.340
18	23.486	47.869	73.486	21.083	44.516	71.083	16.343	38.047	66.343
19	23.488	47.872	73.488	21.086	44.521	71.086	16.345	38.050	66.345
20	23.489	47.874	73.489	21.088	44.524	71.088	16.347	38.053	66.347
21	23.491	47.876	73.491	21.090	44.526	71.090	16.348	38.054	66.348
22	23.491	47.877	73.491	21.091	44.527	71.091	16.349	38.055	66.348
23	23.492	47.878	73.492	21.092	44.529	71.092	16.349	38.056	66.349
24	23.492	47.878	73.492	21.092	44.529	71.092	16.349	38.057	66.349
25	23.493	47.878	73.493	21.093	44.530	71.093	16.350	38.057	66.350
26	23.493	47.879	73.493	21.093	44.530	71.093	16.350	38.057	66.350
27	23.493	47.879	73.493	21.093	44.531	71.093	16.350	38.058	66.350
28	23.493	47.879	73.493	21.093	44.531	71.093	16.350	38.058	66.350
29	23.493	47.879	73.493	21.093	44.531	71.093	16.350	38.058	66.350
30	23.493	47.879	73.493	21.094	44.531	71.094	16.350	38.058	66.350
Actual Solution	(23.493)	(47.879)	(73.493)	(21.094)	(44.531)	(71.094)	(16.350)	(38.058)	(66.350)

Table 5-3 The Results of Example 5-1 by the Method of Successive Displacements

Iteration number	T_1	T_2	T_3	T_4	T_5	T_6	T_7	T_8	T_9
Start	20.313	43.750	70.313	15.625	37.500	65.625	10.983	31.250	60.938
1	21.094	44.727	71.338	17.383	39.746	68.005	13.721	34.851	64.777
2	21.777	45.715	72.180	18.811	41.846	69.701	14.978	36.650	65.650
3	22.382	46.602	72.826	19.801	43.188	70.416	15.675	37.378	66.011
4	22.851	47.216	73.158	20.429	43.860	70.757	16.014	37.721	66.182
5	23.161	47.545	73.325	20.759	44.196	70.926	16.183	37.890	66.266
6	23.326	47.712	73.409	20.926	44.363	71.010	16.266	37.974	66.308
7	23.409	47.796	73.451	21.010	44.447	71.052	16.308	38.016	66.329
8	23.451	47.837	73.472	21.052	44.489	71.073	16.329	38.037	66.340
9	23.472	47.858	73.483	21.073	44.510	71.083	16.340	38.048	66.345
10	23.483	47.869	73.488	21.083	44.521	71.088	16.345	38.053	66.348
11	23.488	47.874	73.491	21.088	44.526	71.091	16.348	38.055	66.349
12	23.491	47.877	73.492	21.091	44.529	71.092	16.349	38.057	66.350
13	23.492	47.878	73.493	21.092	44.530	71.093	16.350	38.057	66.350
14	23.493	47.879	73.493	21.093	44.531	71.093	16.350	38.058	66.350
15	23.493	47.879	73.493	21.093	44.531	71.094	16.350	38.058	66.350
16	23.493	47.879	73.493	21.094	44.531	71.094	16.350	38.058	66.350
Actual Solution	(23.493)	(47.879)	(73.493)	(21.094)	(44.531)	(71.094)	(16.350)	(38.058)	(66.350)

Table 5-4   The Results of Example 5-1 by the Method of Successive Over-Relaxation ($\omega = 1.2$)

Iteration Number	T_1	T_2	T_3	T_4	T_5	T_6	T_7	T_8	T_9
Start	20.313	43.750	70.313	15.625	37.500	65.625	10.938	31.250	60.938
1	21.250	44.969	71.616	17.781	40.387	68.757	14.397	35.967	66.105
2	22.075	46.230	72.673	19.502	43.059	70.799	15.636	37.747	66.218
3	22.804	47.315	73.400	20.549	44.311	71.019	16.237	37.980	66.331
4	23.298	47.840	73.478	21.044	44.503	71.090	16.335	38.055	66.352
5	23.505	47.878	73.495	21.094	44.534	71.096	16.353	38.061	66.352
6	23.490	47.880	73.494	21.094	44.533	71.094	16.351	38.058	66.350
7	23.494	47.880	73.493	21.094	44.532	71.094	16.351	38.058	66.350
8	23.494	47.880	73.493	21.094	44.531	71.094	16.350	38.058	66.350
9	23.493	47.879	73.493	21.094	44.531	71.094	16.350	38.058	66.350
Actual Solution	(23.493)	(47.879)	(73.493)	(21.094)	(44.531)	(71.094)	(16.350)	(38.058)•	(66.350)

extrapolation using two successive displacement steps. In this sense the method of successive over-relaxation can be viewed as an extension of the method of successive displacements discussed previously. If a node point has a present value of $T_{i,j}^{(k-1)}$ and if the method of successive displacements predicts a new value of $\bar{T}_{i,j}^{(k)}$ then the value actually used will be:

$$T_{i,j}^{(k)} = T_{i,j}^{(k-1)} + \omega\ [\bar{T}_{i,j}^{(k)} - T_{i,j}^{(k-1)}]$$

In this expression "i,j" denotes the position of the point in the computational grid, "k" denotes the iteration number in the interation process, and "ω" is called the "relaxation parameter." This parameter must be restricted to values:

$$1 \leqq \omega < 2.$$

The choice for this parameter will determine the speed of convergence. If $\omega = 1$, the process degenerates to the method of successive displacements. Although it is beyond the scope of this text, it is possible to compute an optimum value for the choice of ω. The interested reader may consult Ames [1] for an explanation of the procedure. The alternative to this calculation process is to choose an arbitrary value for ω on the interval $1 \leqq \omega < 2$ and to observe how the convergence progresses.

If the method of successive over-relaxation is applied to Example 5-1 using $\omega = 1.2$, the results are as shown in Table 5-4. Clearly, this method is the best of all of those presented. Although the example problem considered in this section is a two-dimensional problem, the iterative methods are by no means limited to this type of problem. Indeed, a three-dimensional problem can be solved by these same methods if an equivalent three-dimensional computational template is used.

The topic of partial differential equation solutions for propagation problems is now to be considered.

5.7   Parabolic Partial Differential Equations

In order to demonstrate how the finite difference approach can be applied to the solution of parabolic differential equations, a simple example problem is considered.

Example 5-2

Suppose that a uniform cross-section bar of length L is embedded in an insulating medium so that only its left end is exposed to the environment, as shown in the following figure. If the bar is initially at an equilibrium temperature of T = 0 and

117

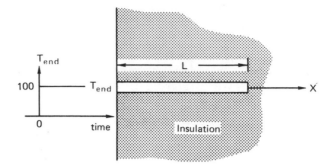

the left end is suddenly subjected to T = 100°C at t = 0 (a step input) find the temperature-vs-time history of various points along the length of the bar.

The partial differential equation that describes the temperature-vs-time and position for this bar will be:

$$K \frac{\partial^2 T}{\partial x^2} = \frac{\partial T}{\partial t} \qquad \begin{array}{c} 0 \leq x \leq L \\ 0 < t \end{array}$$

where K is the thermal diffusivity of the material and depends on the thermal conductivity, the specific heat, and the mass density of the bar. The boundary conditions are:

$$T(0,t) = T_0+ = 100$$

$$\frac{\partial T}{\partial x}(L,t) = 0,$$

and the initial condition is:

$$T(x,0) = T_0 = 0.$$

The numerical solution for this problem can be made more general if the variables are nondimensionalized by defining:

$$\bar{x} = \frac{x}{L}, \quad \bar{t} = \frac{tK}{L^2}, \text{ and } \quad \bar{T} = \frac{T - T_0}{T_0+ - T_0}$$

When this is done, the equation (written without the bars) will be:

$$\frac{\partial^2 T}{\partial x^2} - \frac{\partial T}{\partial t} = 0.$$

The new boundary will be:

$$T(0,t) = 1.0$$

$$\frac{\partial T}{\partial x}(1,t) = 0.0,$$

and the new initial condition will be:

$$T(x,0) = 0.0$$

Finite difference forms for the derivatives may be applied to obtain:

$$\frac{\partial^2 T}{\partial x^2} - \frac{\partial T}{\partial t} = 0 = \frac{T_{i+1,j} - 2T_{i,j} + T_{i-1,j}}{h^2} - \frac{T_{i,j+1} - T_{i,j}}{k}.$$

In this expression the derivative with respect to "x" is a central-difference form, and the derivative with respect to "t" is a forward-difference form. It is interesting to note that the computational grid representing this problem need not be square because the value of "h" need not equal "k". If one lets r = k/h^2, the value of $T_{i,j+1}$ can be found to be:

$$T_{i,j+1} = rT_{i+1,j} + (1 - 2r)T_{i,j} + rT_{i-1,j}.$$

This finite difference equation holds for any interior point and is an explicit formula for finding the temperature at time "t + k" in terms of the temperatures at time "t". This fortunate formulation does not give rise to simultaneous equations; thus iterative techniques are not required. The resulting solution will have errors on the order of h^2 and k^2 since these are the orders of the terms dropped in the finite difference approximations. The appearance of this problem would give the user opportunity to think that any selection for step size can be used for either variable. Unfortunately, this is not true. In the previous chapter, we observed how some differential equation solutions can diverge from the exact solution. The quality describing this behavior was referred to as stability. In the case of partial differential equations, the solutions can diverge and can also oscillate.

In the case of the problem considered in this section, the value of "r" may be used as a test for the stability of the resulting solution. If r $\leq$ ½, the solution will be stable but may oscillate slightly. The mathematical derivation of this fact is presented by Salvadori and Baron [13].

Even though it does have the potential for slight oscillation, the value of r = ½ is a convenient choice because it causes the recursion relationship to become:

$$T_{i,j+1} = 0.5(T_{i+1,j} + T_{i-1,j}).$$

Of course, once the horizontal spacing is chosen for a given r value, the vertical spacing is no longer a free choice. Thus if h = 0.2 and r = ½, then k = 1/50. A grid network reflecting these proportions is presented in the following figure.

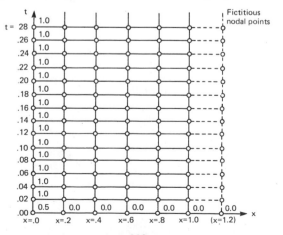

119

Since a discontinuity exists between the boundary condition and the initial condition at the point (0,0), the temperature value assigned at this point is 0.5, which is an average of the two possible values. This procedure is common practice in the solution of problems of this type. The zero slope at the right end of the bar is modeled by means of the addition of a fictitious line of node points at x = 1.2 that have temperature values identical to the points at x = 0.8. Using this grid and the recursion formula, the solution may be carried for as many values of time as desired. The output for this problem is shown in the Table 5-5.

5.8 Hyperbolic Partial Differential Equations

Perhaps the most common engineering form of the hyperbolic partial differential equation is the wave equation. This equation describes various types of oscillatory behavior such as the vibration of strings, the vibration of membranes, or the propagation of acoustic waves in a fluid.

The general form of the wave equation is:

$$C^2 \nabla^2 F = \frac{\partial^2 F}{\partial t^2},$$

where C is a constant. In the case of the free, transverse vibration of a string:

$$C^2 = T/\rho,$$

where T is the tension in the string and ρ is the mass of the string-per-unit-length. For the vibrating string phenomenon, the partial differential equation can be nondimensionalized by a technique similar to that used in the previous section to get:

$$\frac{\partial^2 F}{\partial x^2} = \frac{\partial^2 F}{\partial t^2}.$$

For this equation two boundary conditions and two initial conditions are required to complete the description of the problem. The equation can be written in finite difference form using central differences to get:

$$\frac{F_{i+1,j} - 2F_{i,j} + F_{i-1,j}}{h^2} = \frac{F_{i,j+1} - 2F_{i,j} + F_{i,j-1}}{k^2}$$

If the substitution $r = k/h$ is made, the expression for $F_{i,j+1}$ will be:

$$F_{i,j+1} = r^2 F_{i+1,j} + 2(1-r^2)F_{i,j} + r^2 F_{i-1,j} - F_{i,j-1}.$$

Like the relationship developed for the parabolic equation, this is an expression that allows movement forward in time without the need for iteration. Unlike the previous formula, this recursion relationship requires values of F at both $t = t_j$ and $t = t_{j-1}$ in order to get F at $t = t_{j+1}$. The

Table 5-5 Nondimensional Temperature

time	x_0	x_1	x_2	x_3	x_4	x_5	(x_6)
0.00	0.5000	0.0000	0.0000	0.0000	0.0000	0.0000	0.0000
0.02	1.0000	0.2500	0.0000	0.0000	0.0000	0.0000	0.0000
0.04	1.0000	0.5000	0.1250	0.0000	0.0000	0.0000	0.0000
0.06	1.0000	0.5625	0.2500	0.0625	0.0000	0.0000	0.0000
0.08	1.0000	0.6250	0.3125	0.1250	0.0313	0.0000	0.0313
0.10	1.0000	0.6563	0.3750	0.1719	0.0625	0.0313	0.0625
0.12	1.0000	0.6875	0.4141	0.2188	0.1016	0.0625	0.1016
0.14	1.0000	0.7070	0.4531	0.2578	0.1406	0.1016	0.1406
0.16	1.0000	0.7266	0.4824	0.2969	0.1797	0.1406	0.1797
0.18	1.0000	0.7412	0.5117	0.3311	0.2188	0.1797	0.2188
0.20	1.0000	0.7559	0.5361	0.3652	0.2554	0.2188	0.2554

choice for the aspect ratio of the calculation grid is determined by the value "r", which is also a measure of the stability of the resulting solution. Salvadori and Baron [13] report that for $r > 1$ the finite difference approximation is unstable, for $r = 1$ the approximation is stable and the finite difference solution agrees with the exact solution, and for $r < 1$ the solution is stable but exhibits decreasing accuracy as r decreases. The choice of $r = 1$ is fortunate because it also simplifies the explicit recursion formula to:

$$F_{i,j+1} = F_{i+1,j} + F_{i-1,j} - F_{i,j-1}.$$

Using this formula and the boundary/initial conditions, the solution can be constructed in a manner similar to that demonstrated in the previous section.

5.9 Numerical Solution of Partial Differential Equation Problems by the Finite Element Method

The finite element method of describing continuous systems was first introduced in the mid-1950s and has since become an extremely useful engineering technique. The finite element method has found applications in a variety of fields including stress analysis, fluid dynamics, and field theory. As a quantitative measure of the acceptance of this engineering tool, Norrie and de Vries [9] list over 7000 references to this method in the recent engineering and scientific literature. Although the finite element method is applied to similar problems as the difference method, the two methods are quite different in concept. The finite difference method starts with the differential equation and approximates the derivatives in this equation with discrete steps. The finite element method has its mathematical basis in an inverse approach to the calculus of variations. The governing differential equation and associated boundary conditions are transformed into an extremum problem which is then solved directly. In this sense the finite element method is an implicit piecewise application of the Ritz method. The finite element method utilizes a piecewise substitute model for the physical problem being considered. In this sense the finite element method allows the engineering user to exploit his or her intuitive grasp of the problem being considered. Although an adequate theoretical treatment of the details of the finite element method would require a full textbook and is well beyond the scope of this presentation, a basic description of this important method is now given. For a description more involved than the one here presented, the interested reader may consult the works of Cook [2] or Zienkiewicz and Cheung [16] listed at the end of this chapter. For ease of discussion, the finite element method is here described in terms of structural analysis, although this discussion could easily apply to other areas of engineering application.

The basic steps involved in performing a finite element analysis are presented in Figure 5-8. The first step is to divide the body into small simple elements connected at points called nodes. Considerable versatility exists in the selection and placement of these simple elements, because the engineer is free to exercise judgment in choosing size, type, shape, and orientation. Frequently these elements are triangles or quadrilaterals in the plane or tetrahedrons or hexahedrons in three-dimensional problems. In a location on the physical problem where more detailed information is desired, the engineer might use many small elements. If material or physical properties of the part change at a point or along a seam, the engineer may wish to change the form, size, or orientation of the elements at these special locations. Figure 5-9 shows how a square plate with an elliptical hole subjected to a uniform stress field might be modeled by 26 triangular finite elements. Because the plate has twofold symmetry, it is only necessary to model a quarter segment. Notice how the elements have deliberately been made smaller in the area near to the elliptical hole. This will allow the engineer to obtain more detailed information in the location where the stress levels are rapidly changing. As is indicated in Figure 5-9, it is customary to number both the elements and the nodes. This procedure enables the input to a finite element program to be accomplished in an efficient manner. The uniform stress field in this illustration has been modeled by the application of concentrated force loads on nodes 21, 20, and 19.

The next step in the finite element analysis process is to assume some simple interpolation scheme to represent the displacement at any internal point within the element in terms of the node coordinate values. Often this assumed displacement function will be a simple polynomial. Later in the solution process, these piecewise polynomials will be used to interpolate displacement values within the elements.

Next, the elastic relationships between node loads and node deflections are written for each element. The engineering concept of matrix treatment of influence coefficients can be very useful in this stage of the analysis. Once these load-vs-deflections are known, the load-vs-deflection equations for the overall physical system can be constructed by requiring that the node deflection values for adjacent elements be equal and that forces at nodes must add up to the applied external loads at these locations. The result will be a linear system of equations that is of the form:

$$[K]\{d\} = \{R\}$$

where [K] is known as the system stiffness matrix, $\{d\}$ is known as the

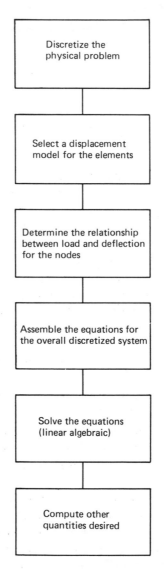

Figure 5-8   Steps in the finite element analysis method.

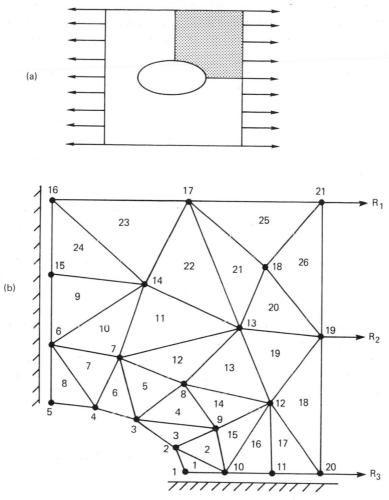

(a)

(b)

Figure 5-9 Example of a finite element representation of a square plate with an elliptic hole subjected to a uniform stress field.

system displacement vector, and {R} is known as the system force vector. This system of equations will be somewhat sparse owing to the fact that not every node is contained in every element. In a general displacement problem, if there are "m" nodes, each node will be able to have "n" independent displacement coordinates (perhaps x and y coordinates in two-dimensions). Thus the stiffness matrix will be of size nm x nm, the system node displacement vector will be of size nm, and the system force vector will be of size nm. Some displacement values will be known by virtue of the known constraints. For example, in Figure 5-9(b) the horizontal displacement of nodes 5, 6, 15, and 16 is zero and the vertical displacement of nodes 1, 10, 11, and 20 must be zero. Known displacement values may be eliminated from the equation system to reduce its order. Alternatively, the appropriate stiffness term (at a particular point along the diagonal of the stiffness matrix) may be set to some large value that is large relative to all other stiffness terms. This procedure has the advantage that it preserves the original form of the equation system and is often used in packaged finite element programs. Because the equation system is sparse, a method such as successive over-relaxation is used to find the displacement values for each node. Once the displacement behavior of the physical system has been approximated numerically, the traditional equations of the theory of elasticity can be applied to find the strains or the stresses if desired.

It is obvious in this formulation that matrix methods can be applied to assist with the organization of the problem assembly and solution. For this reason most finite element texts begin with a presentation of matrix fundamentals. Because it lends itself to a systematic formulation, the finite element method has been generalized to the point where general-purpose computer programs have been developed for its implementation. Table 5-6 summarizes some of the more familiar computer programs that are available for finite element analysis. Although these programs are easy to apply, they can consume considerable computer time to complete a run and frequently require large amounts of computer storage. Nevertheless, they are extremely helpful to the engineering problem solver who faces a complex analysis problem.

Table 5-6   General Purpose Finite Element Programs

Program Name	Developer
ANSYS	Swanson Analysis Systems, Inc. 870 Pine View Drive Elizabeth, PA 15037
ASAS	Atkins Research and Development Ashley Road Epsom, Surrey, England
ASKA	ASKA-Group Pfaffenwaldring 27 Stuttgart-80, West Germany
BERSAFE	CEGB Berkeley Nuclear Laboratory Berkeley, Glos., U.K.

(continued)

Program Name	Developer
BOSOR4	Dr. David Bushnell, Dept. 5233 Lockheed Missiles & Space Co. 3251 Hanover St. Palo Alto, CA 94304
DYNAL	A. Y. Cheung Ontario Hydro 620 University Ave. Toronto 2, Ontario, Canada
EASE	Engineering Analysis Systems 1611 S. Pacific Coast Highway Redondo Beach, CA 90277
MARC	Analysis Corporation 105 Medway St. Providence, RI 02960
NASTRAN	C. W. McCormick Macneal-Schwendler Corp. 7422 N. Figueroa St. Los Angeles, CA 90041
NEPSAP	P. Sharifi Lockheed Dept. 81-12, Bldg. 154 P. O. Box 504 Sunnyvale, CA 94087
NONLIN2	J. C. Anderson Sargent & Lundy 140 S. Dearborn Chicago, IL 60603
NONSAP	K. Bathe NISE, 729 Davis Hall Univ. of California Berkeley, CA 94720
SAPIV	K. Bathe NISE, 729 Davis Hall Univ. of California Berkeley, CA 94720
STARDYNE	R. Rosen Mechanics Research Inc. 9841 Airport Bld. Los Angeles, CA 90045
STRUDL II	ICES User's Group Inc. P. O. Box 8243 Cranston, RI 02920
SUPERB	Structural Dynamics Research Corp. 5729 Dragon Way Cincinnati, OH 45227

5.10 General Considerations in the Solution of Partial Differential Equations

Because each partial differential equation is somewhat unique, and because the boundary conditions associated with a given problem make it even more unique, it is difficult to state a large number of guidelines that are universally useful in the solution of partial differential equations. Nevertheless, a few important considerations should be kept in mind in the solution process:

1. Consider the Desired Accuracy of the Solution. If high accuracy is desired for the solution of a given partial differential equation, it may be necessary to use a very small grid or very small element sizes. The approximations used in the finite difference formulations of this chapter have errors on the order of h^2.

2. Consider the Shape of the Problem Being Considered. Quite frequently, knowledge of symmetry can be applied to reduce the number of node points in a given problem by a factor of 2 or 4. This can result in a significant saving in computer time and computer storage.

3. Consider the Starting Values. In iterative methods, the accuracy of the initial approximation to the solution will influence the speed of convergence. In the absence of good initial approximations, it may prove worthwhile to do the problem in two or more stages. In the first stage, an extremely coarse grid or extremely large finite elements are used to generate good starting values for a later solution attempt using a much finer grid or element mesh.

4. Consider What Type of Method Best Suits the Problem Situation. A question always arises as to whether the finite element method or the finite difference method is best to use. The answer to this question lies in a consideration of the problem itself and in a consideration of the computer being used to solve the problem. Table 5-7 presents a comparison of the basic features of the two methods so that the user may make a choice based on a factual consideration of their relative performance characteristics.

Table 5-7 A Comparison of the Finite Element Method and the Finite Difference Method

Finite Element Method	Finite Difference Method
An approximate method for solving problems described by differential equations.	An approximate method for solving differential equations.
Involves the solution of large, sparse systems of equations.	Involves the solution of large, sparse systems of equations.

(Continued)

Finite Element Method	Finite Difference Method
The setup can be rather involved with many inputs and outputs.	The setup is rather simple with few inputs and few outputs.
Considerable variety of choice exists in the selection of element type, shape, and size.	Regular mesh patterns are usually required. It is difficult to obtain economical mesh grading.
Structure closely resembles the actual physical problem.	Structure does not usually resemble the actual problem.
Can handle nonhomogeneous and anisotropic problems.	Not well suited for nonhomogeneous and anisotropic problems.
Can be generalized and thus many existing programs are available.	Each problem setup is unique.
Requires skill and judgment to set up.	Solution is straightforward once one has the differential equation.
A specific solution applies only to a specific problem--solutions cannot be generalized.	A specific solution applies only to a specific problem--solutions cannot be generalized.
Boundary conditions are easy to represent.	Unusual boundary conditions are difficult to represent.

Problems

5.1 Laplace's equation for rectangular coordinates in three dimensions is:

$$\nabla^2\phi = 0 = \frac{\partial^2\phi}{\partial x^2} + \frac{\partial^2\phi}{\partial y^2} + \frac{\partial^2\phi}{\partial z^2}$$

What would the computational template for this equation look like?

5.2 How would the computational template for the biharmonic operator be modified to account for the irregular boundary shown in the following figure:

129

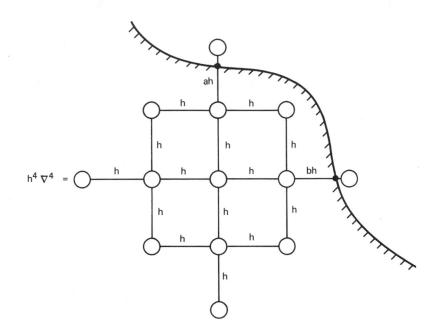

$h^4 \nabla^4 =$

5.3 The transverse deflection w of a thin plate subjected to a uniform pressure p is expressed as:

$$\nabla^4 w = \frac{p}{D},$$

where D is the flexural rigidity and is found from:

$$D = \frac{E\,t^3}{12(1-v^2)},$$

where:

E = the elastic modulus

t = the thickness of the plate

v = Poisson's ratio for the material.

A steel plate (E = 210 x 10^9 N/m^2 v = 0.3) of thickness t = 0.005 m is subjected to a uniform pressure p = 100N/m^2 and is clamped along all edges, as shown in the following figure. Write and run a computer program that will determine the deflection of this plate.

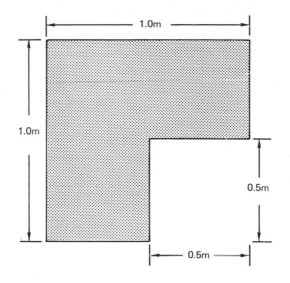

5.4 The transverse deflection of a membrane will be described differently than that of a plate, because it cannot have bending or shear stresses. The transverse deflection w of a membrane will be described by:

$$\frac{\partial^2 w}{\partial x^2} + \frac{\partial^2 w}{\partial y^2} = -\frac{p}{T},$$

where p is the pressure on the membrane and T is the tension-per-unit length on the membrane. For relatively small deflections, it may be assumed that the membrane tension is a constant. If $T = 10^7$ N/m, $p = 10^4$ N/m^2, and the geometry shown in Problem 5.3 is used, write a computer program that will predict the deflections for the membrane.

5.5 Two tubes carrying a hot liquid at 90°C are embedded in a square cross-section duct of insulating material as shown. If the outer temperature of the material is found to be 30°C and the effect of contact conductance is neglected, find and plot isotherms corresponding to T = 40°C, 50°C, 60°C, 70°C, and 80°C.

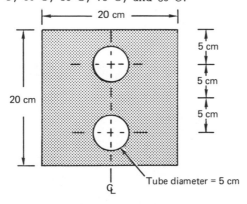

131

5.6 The Navier-Stokes equation for steady, viscous flow in a conduit of arbitrary constant cross section takes the form:

$$\frac{\partial^2 w}{\partial x^2} + \frac{\partial^2 w}{\partial y^2} = \frac{1}{\mu} \frac{dp}{dz},$$

where w is the velocity (being zero at all points on the conduit boundary), μ is the coefficient of viscosity, and dp/dz is the constant pressure drop along the conduit. If dp/dz = -5000 N/m^2/m and μ = 1.5 x 10^{-4} N sec/m^2, write a computer program that will predict the velocity of flow in the conduit shown in the following figure.

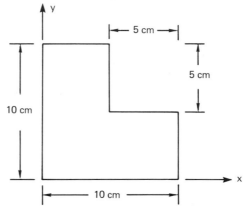

5.7 Rework Example 5-2 if the right-hand end of the uniform cross-section bar is touching an infinite energy sink at 0°C. If the bar is made of steel, how long would it take for the bar to reach a steady state temperature distribution if the length is 10 cm?

5.8 The longitudinal vibration of a beam will be described by the wave equation:

$$\frac{\partial^2 u}{\partial x^2} = \frac{\rho}{E} \frac{\partial^2 u}{\partial t^2},$$

where u is the longitudinal displacement, E is the elastic modulus for the material, and ρ is the density of the material. If a pinned-pinned beam of length L is grasped at the center, is held in a deflected position u*, and is then released from rest, write a computer program that will describe the displacement of the beam. In order to make this computer solution more general, it is often convenient to let:

$$\bar{x} = x/L$$

$$\bar{t} = t[E/\rho]^{\frac{1}{2}}/L^2$$

$$\bar{u} = u/u^*.$$

This will cause the wave equation to become:

$$\frac{\partial^2 \bar{u}}{\partial \bar{x}^2} = \frac{\partial^2 \bar{u}}{\partial \bar{t}^2}$$

5.9 A 10 cm x 10 cm square plate of constant thickness has a 1 cm diameter hole in its center as shown. The plate is loaded with a uniform tensile load of 10,000 Pa. Formulate this problem in finite element form and, if a general purpose finite element program is available to you, use the program to find the location of the maximum stress that occurs in the part. What would you predict for the stress concentration factor for this geometry?

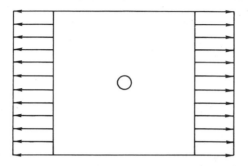

5.10 If the plate in Problem 5.9 were loaded in pure bending rather than in pure tension, what would be the resulting maximum stress location, and what would be the stress concentration factor?

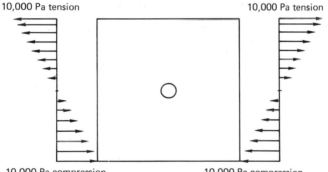

References

1. Ames, W. F. Numerical Methods for Partial Differential Equations. New York: Barnes & Noble, 1969.

2. Cook, R. D. Concepts and Applications of Finite Element Analysis. New York: John Wiley & Sons, 1974.

3. Greenspan, D. Introduction to Partial Differential Equations. New York: McGraw-Hill Book Co., 1961.

4. Killer, H. B. "On Some Iterative Methods for Solving Elliptic Difference Equations." Quarterly of Applied Mathematics 16 (Oct. 1958).

5. Ketter, R. L., and Prawel, S. P. <u>Modern</u> <u>Methods</u> <u>of</u> <u>Engineering</u> <u>Computation</u>. New York: McGraw-Hill Book Co., 1969.

6. Liebmann, H. "Dieangenaherte Ermittlung Harmonischer Funktionen und konformer abbildung." Sitzber. Math. Physik, Kl. Bayer. Akad. Wiss. (1918): 385.

7. Machine Design. <u>A</u> <u>Basic</u> <u>Course</u> <u>in</u> <u>Numerical</u> <u>Methods</u> <u>and</u> <u>FORTRAN</u> <u>Programming</u>. Reprinted from Machine Design, October 26, 1967 through June 20, 1968.

8. Moon, P., and Spencer, D. E. <u>Field</u> <u>Theory</u> <u>for</u> <u>Engineers</u>. New York: Van Nostrand Co., 1971.

9. Norrie, D. and de Vries, G. <u>Finite</u> <u>Element</u> <u>Bibliography</u>. New York: Plenum Press, 1976.

10. Pall, G. A. <u>Introduction</u> <u>to</u> <u>Scientific</u> <u>Computing</u>. New York: Appleton-Century-Crofts, 1971.

11. Pilkey, W., Saczalski, K., and Schaeffer, H. <u>Structural</u> <u>Mechanics</u> <u>Computer</u> <u>Programs-Surveys</u>, <u>Assessments</u>, <u>and</u> <u>Availability</u>. University of Virginia Press, Charlottesville, 1974.

12. Ralston, A., and Wilf, H. S. <u>Mathematical</u> <u>Methods</u> <u>for</u> <u>Digital</u> <u>Computers</u>, New York: John Wiley & Sons, 1967.

13. Salvadori, M. G., and Baron, M. L. <u>Numerical</u> <u>Methods</u> <u>in</u> <u>Engineering</u>. Englewood Cliffs, N.J.: Prentice-Hall, Inc., 1961.

14. Southwell, R. V. <u>Relaxation</u> <u>Methods</u> <u>in</u> <u>Theoretical</u> <u>Physics</u>. London: Oxford University Press, 1946.

15. Young, D. M. <u>Trans</u>. <u>of</u> <u>the</u> <u>Am</u>. <u>Math</u>. <u>Soc</u>. 16, (1968): 209.

16. Zienkiewicz, O. C., and Cheung, Y. K. <u>The</u> <u>Finite</u> <u>Element</u> <u>Method</u> <u>in</u> <u>Structural</u> <u>and</u> <u>Continuum</u> <u>Mechanics</u>. New York: McGraw-Hill Book Co., 1967.

6 OPTIMIZATION—PART I

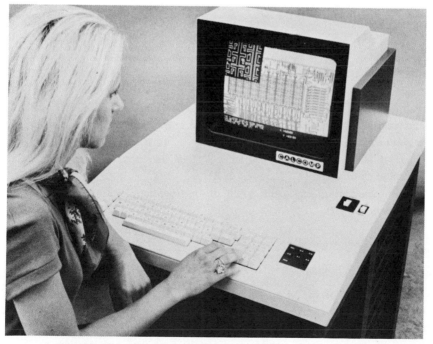

Through the medium of computer graphics the engineer can greatly expand his or her ability to visualize and to manipulate engineering ideas. [Photograph courtesy of California Computer Products, Inc. (CalComp).]

6.1   Introduction

Optimization provides a logical method for the selection of the best choice from among all possible designs that are available. In recent years this topic has received considerable treatment in the literature, and a number of excellent computer-aided optimization algorithms are available. In this chapter we first discuss the fundamental forms of optimization problems. Then we discuss the rationale upon which the basic algorithms are based. Finally, we study the most well-known algorithms in order to compare and contrast their strengths and weaknesses as design aids.

6.2   Fundamentals of Optimization

The term "optimization," as it is used in the design literature, is defined as the process or rationale that a designer uses to achieve an improved solution. Although it is desirable to have the very best or "optimum" solution to a problem, engineers usually must settle for improvement rather than perfection in their designs. For this reason we speak of optimization as the process of movement toward improvement rather than the achievement of perfection.

If we consider a system defined in terms of "m" equations and "n" unknowns, we can identify three fundamental types of problems. If m = n, the problem is said to be algebraic and usually has at least one solution. If m > n, the problem is said to be overconstrained and is impossible to solve in general. If m < n, the problem is underconstrained and many usable solutions exist that satisfy the requirements. In design, this third class of problems occurs most frequently. When it does, the designer must exercise his or her creative talents to select additional constraints or criteria upon which to base the design choice. Clearly, the product or process that performs better than its competitors will achieve the most success in the marketplace. Thus the importance of optimization in design is obvious.

In order to discuss the topic of optimization, we need a few basic definitions.

Design Variables

The term "design variable" is used to describe the individual elements in a group of independent, variable parameters that uniquely and completely define the design problem being considered. The design variables are unknown values to be solved in the optimization process.

136

They take the form of any of the fundamental or derived units that can be used to quantify engineering systems. Thus, for example, they may represent unknown values of length, mass, time, and temperature. The actual number of design variables determines the complexity and versatility of a given design problem. We generally identify the number of design variables with the integer "n" and the design variables become "x" values having subscripts from 1 to n. Thus the "n" design variables in a given problem are:

$$x_1, \ x_2, \ x_3 \ ,.., \ x_n.$$

Merit Function

The merit function is an equation or expression that the designer desires to maximize or minimize. It provides a quantitative means for evaluating and comparing the relative quality of two competing designs. Mathematically, the merit function defines an "n + 1" dimensional surface. The value of the merit function will depend on the values of the design variables:

$$M = M(x_1, \ x_2, \ \ldots, \ x_n).$$

Examples of common engineering merit functions to be maximized or minimized are cost, weight, strength, size, and efficiency. If there is only one design variable, the merit function can be plotted as graph in two dimensions as shown in Figure 6-1. If there are two design variables, the

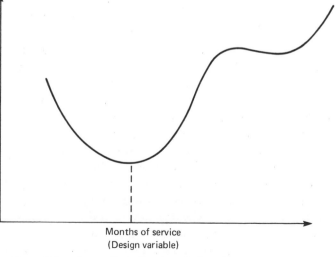

Months of service
(Design variable)

Figure 6-1   A one-dimensional merit function.

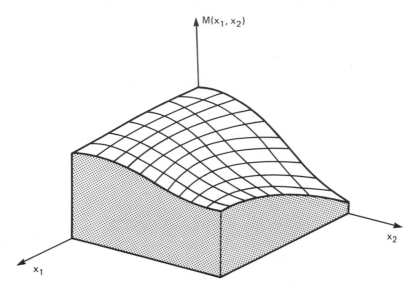

Figure 6-2 A two-dimensional merit function.

merit surface can be plotted as a three-dimensional surface, as shown in Figure 6-2. The merit surface for three or more design variables is called a "hypersurface" and cannot be visualized in the traditional sense. The physical and mathematical characteristics of the merit surface are of strong importance in the optimization process, because the nature of this surface influences the selection of the most efficient algorithm.

The merit function, as we have defined it, can assume some unusually challenging forms in a given design situation. For example, the merit function sometimes cannot be expressed in closed mathematical form or may involve a piecewise continuous function. Evaluation of the merit function may sometimes require the use of a table of engineering information such as the steam tables or may rely on data gathered in an experiment. For some types of design problems, the design variables can only assume integer values rather than real values. Examples of this type of design variable include such things as the number of teeth on a gear or the number of bolts on a flange. Occasionally, design variables can only assume a value of yes or no rather than a number. Qualitative factors such as customer satisfaction, safety, and esthetic appeal are difficult to use in optimization schemes because they are cumbersome to quantify numerically. Regardless of the form it takes, however, the merit function must be a unique function of the design variables.

Some types of optimization problems can be formulated in terms of more than one measure of merit. Sometimes these two may even be in conflict. An example of this situation occurs in aircraft design, where it is desired to maximize strength, minimize weight, and minimize cost. Whenever the possibility for more than one measure of merit exists, the designer must establish priorities and assign weighting values to each measure of merit. This process results in what is called a "tradeoff" function and provides a single composite merit value to be used in the optimization process.

Minimization and Maximization

Some optimization algorithms are set up to search for maxima, and others are set up to search for minima. Regardless of the type of extremum problem being solved, a general algorithm may be used because a minimization problem can be converted into a maximization problem by simply changing the sign of the merit value. This situation is illustrated in Figure 6-3.

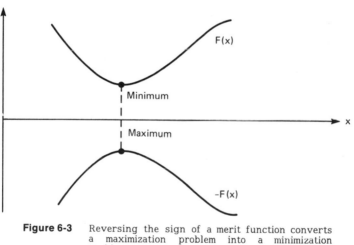

Figure 6-3   Reversing the sign of a merit function converts a maximization problem into a minimization problem.

Design Space

The total domain defined by all of the "n" design variables is called the design space. This space is not as large as one might think, because it is usually limited by constraint bounds imposed by the reality of the problem. Indeed, it is possible to constrain a problem so much that no usable design exists. The two categories of constraints are called functional constraints and regional constraints.

Functional Constraints

Functional constraints are functional relationships of the design variables that must be satisfied in the design solution. These arise due to such things as the laws of nature, economics, law, taste, and available materials. In general, there may be any number of these of the form:

$$C_1(x_1, x_2, \ldots, x_n) = 0$$
$$C_2(x_1, x_2, \ldots, x_n) = 0$$
$$\vdots$$
$$C_j(x_1, x_2, \ldots, x_n) = 0$$

If any of these functional constraints can be used to solve for one of the design variables as a function of all others, this new expression can be used to eliminate that design variable in the optimization process. In this way the degree of dimensionality of the problem is reduced. This procedure is desirable because it generally reduces the complexity of the problem.

Regional Constraints

Regional constraints are special types of functional constraints that take the form of inequalities. In the most general case, any number of these may exist in the form:

$$z_1 \leq r_1 (x_1, x_2, \ldots, x_n) \leq Z_1$$
$$z_2 \leq r_2 (x_1, x_2, \ldots, x_n) \leq Z_2$$
$$\vdots$$
$$z_k \leq r_k (x_1, x_2, \ldots, x_n) \leq Z_k$$

It should be noted at this point that often an optimum value of a merit function does not occur where the gradient of the merit surface is zero. The best design selection often occurs along one of the constraint boundaries.

Local Optimum

The local optimum is a point in the design space that is higher than all other points within its immediate vicinity. Figure 6-4 illustrates a one-dimensional merit function that has two local optima. Frequently, a design space will have many local optima. The designer must not fall prey to selecting the first optimum value he or she finds.

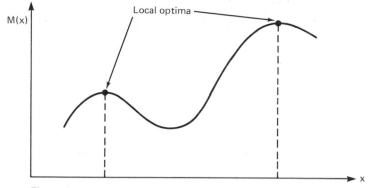

Figure 6-4   A general merit function can have more than one local optimum.

Global Optimum

The global optimum is the optimum design within the total allowable design space. It is the best of all local optima and is the design choice that the designer ultimately seeks to find. It is possible for several equal global optima to occur at two or more places in the design space. The formulation of optimization problems can best be illustrated by means of an example problem.

Example 6-1

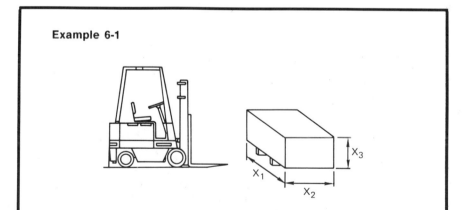

It is desired to design a closed rectangular container to hold 1.0 m³ of a loose fiber product. It is desired to utilize minimum material (i.e., minimum surface area assuming that the same wall thickness is to be used for every surface of the container) for reasons of cost. In order to allow the container to be carried conveniently by a lift truck, the width of the container must be no smaller than 1.5 m.

Formulate this problem in a form suitable for treatment by an optimization algorithm. For this problem:

141

The design variables are: x_1, x_2, and x_3.

The merit function to be minimized is the surface area:

$$A = 2(x_1 x_2 + x_2 x_3 + x_1 x_3) \qquad m^2.$$

The functional constraint is:

$$Volume = 1.0 \ m^3 = x_1 x_2 x_3.$$

The regional constraint is:

$$1.5 \leqq x_1.$$

The careful designer will notice that the degree of dimensionality of this problem can be reduced by virtue of the simplicity of the functional constraint. Because

$$x_3 = \frac{1.}{x_1 x_2},$$

the need for the design value x_3 can be eliminated to get a new problem formulation.

The new problem formulation is:

The design variables are: x_1 and x_2.

The merit function to be minimized is:

$$A = 2(x_1 x_2 + \frac{1}{x_1} + \frac{1}{x_2}) \qquad m^2$$

The functional constraint is: none

The regional constraint is: $1.5 \leqq x_1$.

Once the problem is formulated in this standard form, it is ready to be solved by whatever method the engineer chooses to use. The first impulse of the engineer might be to apply a traditional calculus approach and set:

$$\frac{\partial A}{\partial x_1} = \frac{\partial A}{\partial x_2} = 0.$$

This method will yield $x_1 = x_2 = x_3 = 1.0$ m. Unfortunately this solution violates the regional constraint and is therefore not an acceptable design solution. This example serves to illustrate an important fact about optimization — namely, that due to the constraints on a given problem, the optimum solution may occur at a point other than where the local gradient is zero.

The complete solution to this example problem is possible by the methods presented in the next chapter.

6.3   One-Dimensional Search Techniques

The search for extremes can be compared to the process of finding the deepest point in a lake by means of a series of soundings taken from a boat using a string and a weight. For each sounding new information is gained. If a new depth is larger than previous trials, useful information is gained. If, on the other hand, the new depth is less than previous measurements, the new value is of no use and represents wasted effort. In search methods it is desired to reach the extremum as quickly as possible with a minimum of wasted effort. In this section one-dimensional search techniques to facilitate this process are discussed. In this treatment it is assumed that the merit functions being investigated are "unimodal." This means that they have only a single peak in the interval of interest. Thus, as we make merit evaluations by slowly increasing the design variable, each successive value is progressively larger until we reach the peak. Once past the peak, each successive merit value is progressively less than the previous one. Actually, this limit on the merit surface is not as restrictive as one might think, because many problems in engineering exhibit this type of "single-peak" behavior.

The problem of one-dimensional optimization can be viewed within the following framework. A design variable x must have values between some lower and upper bound so that

$$a \leqq x \leqq b.$$

As the problem beings, nothing is known about the merit function except that it is unimodal. The "interval of uncertainty" is defined as the interval in which the optimum must lie. At the start of the optimization process, this interval is the total length from a to b, as shown in Figure 6-5.

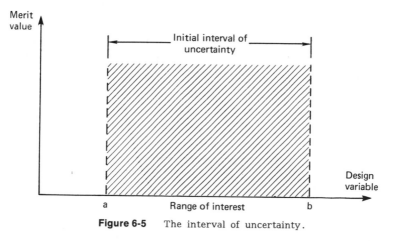

Figure 6-5   The interval of uncertainty.

If two evaluations are made somewhere within the allowable range (say at x_1 and x_2 to get M_1 and M_2), the interval of uncertainty is reduced as shown in Figure 6-6. Several techniques for systematically reducing this interval exist. These are explored in the following subsections.

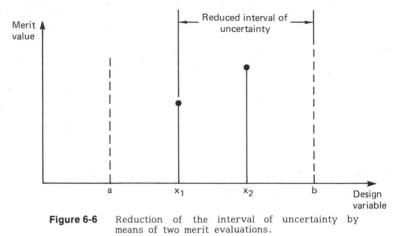

Figure 6-6   Reduction of the interval of uncertainty by means of two merit evaluations.

Total Search

Perhaps the most obvious technique to use in reducing the interval of uncertainty for a one-dimensional unimodal problem is to divide the total interval from a to b into a lattice of equally spaced functional evaluations, as shown in Figure 6-7. As a result of this process, the interval of un-

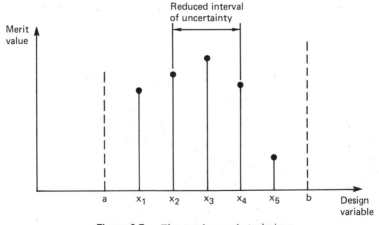

Figure 6-7   The total search technique.

certainty is reduced to a value equal to two lattice spacings. It is customary to speak of the fractional reduction of the interval of uncertainty "f". In this case, for "N" lattice evaluations, the spacing is:

$$\frac{1}{N+1},$$

and the fractional reduction of the interval of uncertainty is:

$$f = \frac{2}{N + 1}.$$

To achieve a reduction of f = 0.01 requires N = 199 evaluations, and to achieve a reduction of f = 0.001 requires N = 1999. Clearly, for this method the efficiency of effort becomes poor as the desired size of the interval of uncertainty gets small. As a logical alternative to achieve f = 0.01, it would be better to expend 19 evaluations to get f = 0.1 and then to expend an additional 19 evaluations on this new, smaller interval of uncertainty to achieve f = 0.01 in 38 evaluations rather than in 199. Thus, with a bit of care, the efficiency of the search can be improved.

Interval Halving

If we apply this logic but allow the number of evalutions in a given subsearch to be a variable, even more efficiency may be obtained. For "N" evaluations accomplished on "I" telescoping subsets, the final degree of reduction for the interval of uncertainty is:

$$f = \left[\frac{2}{N + 1} \right]^{I}.$$

The total number of merit evaluations "J" expended in this search is:

$$J = N*I.$$

It is desired to find the optimum "N" to minimize "J" for a given value of "f". Using I = J/N, we can solve for "J" using the expression for interval of uncertainty. This expression is:

$$J = \frac{N \ln \left[\frac{1}{f} \right]}{\ln \left[\frac{N + 1}{2} \right]}.$$

If the value of J is plotted as shown in Figure 6-8, the minimum value is

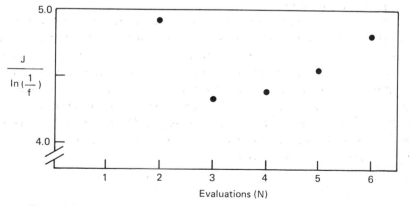

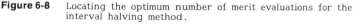

Figure 6-8   Locating the optimum number of merit evaluations for the interval halving method.

145

observed to lie somewhere close to 3. Because the number of evaluations must always be an integer, the value N = 3 will be used as the optimum. For this choice, f = ½ for each subsearch. Since the interval of uncertainty is reduced by one-half, this technique is called interval halving. Figure 6-9 illustrates how the initial three evaluations reduce the interval

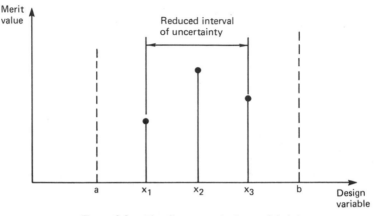

Figure 6-9 The first step in interval halving.

of uncertainty by one-half. Notice, however, that on this new interval the central merit value is already known. Thus, to complete the next subsearch, only two additional evaluations are required rather than three. This saving continues as the algorithm proceeds. In general, the fractional reduction in uncertainty for N evaluations is:

$$f = \frac{1}{2^{\left[\frac{N-1}{2}\right]}} \quad \text{for } N \geq 3.$$

Dichotomous Search

Throughout previous discussions it has been required that the merit evaluations be equally spaced on the interval. If this restriction is lifted, it is possible to achieve a greater efficiency in the search. It has been shown that two merit evaluations on an interval will provide a reduction in the interval of uncertainty. Suppose that these trials are to be placed at special locations arranged so as to achieve the smallest interval of uncertainty as a result of their spacing. Figure 6-10 illustrates the notation used for this scheme. If the merit value of x_1 is greater than that of x_2, then the new interval of uncertainty is:

$$Z_1 = z_1 + z_2.$$

146

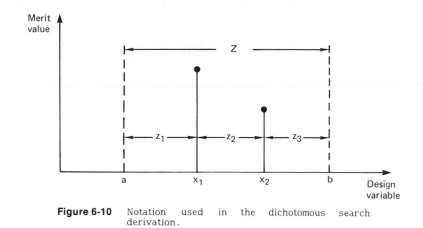

Figure 6-10 Notation used in the dichotomous search derivation.

If the opposite is true, then the new interval of uncertainty is:

$$Z_2 = z_2 + z_3.$$

It is desired to minimize both Z_1 and Z_2 subject to the equality constraint $z_1 + z_2 + z_3 = Z$ and the regional constraints:

$$0 < z_1$$

$$0 < z_2$$

$$0 < z_3.$$

The equality constraint can be used to eliminate z_2 to get:

$$Z - z_3 = Minimum$$

$$Z - z_1 = Minimum$$

Since Z is fixed, the larger z_3 and z_1 become, the smaller these equations will be. Thus the optimum is:

$$z_1 = z_3 = 0.5*Z.$$

But this would give $z_2 = 0$. Since this result violates one of the regional constraints, z_2 must be chosen to be some very small, nonzero value "ε". A value $\varepsilon/2$ will be subtracted from z_1 and z_3 to achieve compatibility. For this choice the interval of uncertainty is reduced to:

$$f = \frac{1}{2} + \frac{\varepsilon}{2}$$

for the first pair of closely spaced evaluations as shown in Figure 6-11. In the limit, as $\varepsilon \to 0$, this uncertainty approaches $f = \frac{1}{2}$. This

147

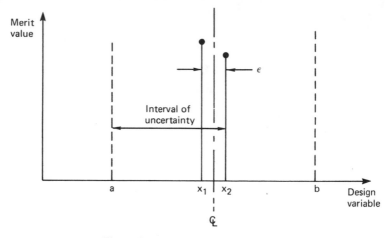

Figure 6-11   The dichotomous search.

dichotomous search then proceeds just as we did with the interval halving method. Notice, however, that the same degree of reduction of the interval of uncertainty has been achieved with one less merit evaluation.

The Golden Section Search

For any three merit evaluations on an interval of uncertainty, two of these will be useful for later evaluations while one will not provide information of further use. It is the purpose of the golden section search to use a nonuniform spacing of merit evaluations arranged so that every single evaluation provides new and useful information. The basis for this scheme is as follows. In dividing an interval of uncertainty into two unequal parts, the ratio of the larger of the two segments to the total length of the interval should be the same as the ratio of the smaller to the larger segment. Thus, if one considers an interval of uncertainty as shown in Figure 6-12, consisting of a length Z composed of two segments

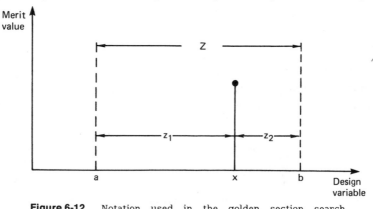

Figure 6-12   Notation used in the golden section search derivation.

148

z_1 and z_2, the golden section theory requires that

$$\frac{z_1}{Z} = \frac{z_2}{z_1}$$

and that

$$z_1 + z_2 = Z.$$

The first equation gives:

$$z_1{}^2 = Zz_2.$$

Substituting for Z from the second equation and dividing through by $z_1{}^2$ gives:

$$1 = \left(\frac{z_2}{z_1}\right)^2 + \left(\frac{z_2}{z_1}\right)$$

This quadratic can be solved for the ratio (z_2/z_1). The positive root is:

$$\frac{z_2}{z_1} = \frac{-1 + \sqrt{5}}{2} = 0.618033989.$$

If the two evaluations on an interval are placed with this fractional spacing from either end, the result is as shown in Figure 6-13. Using these two

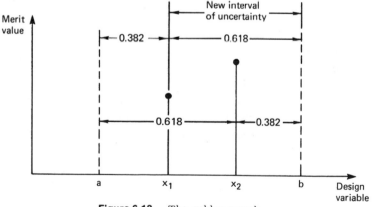

Figure 6-13 The golden search.

evaluations, the interval of uncertainty will be reduced to a length of 0.618 times the previous interval of uncertainty. Although at this stage the method is not as good as interval halving by the dichotomous search, the true advantage is revealed as the new interval is further divided. Here it becomes obvious, by virtue of the golden ratio, that one of the two internal evaluations necessary for the next step is already available.

Thus only one additional "golden-spaced" evalauation is required to reduce the uncertainty by another 0.618 fraction.

The golden section search surpasses the efficiency of the dichotomous search for n > 2, because every single evaluation provides an additional reduction of 0.618 for the interval of uncertainty. For "N" evaluations the final size for the fractional reduction in the interval of uncertainty is:

$$f = (0.618033989)^{N-1}.$$

The golden section search points out an interesting principle. In order to achieve the greatest reduction in subsequent intervals of uncerainty, the merit evaluations should be symmetrically placed about the centerline of the interval of uncertainty. When this is done and each new merit evaluation provides an additional reduction in the interval of uncertainty, the following mathematical expression applies:

$$Z_{I-2} = Z_{I-1} + Z_I \qquad 1 < I < N \quad ,$$

where Z_J represents the length of the interval of uncertainty after the Jth trial evaluation. It should be noted that other symmetrical techniques besides the golden section search can and do exist that satisfy this relationship.

Fibonacci Search

Although the golden section search works quite well, it is obviously not optimum for a given number of evaluations. For example, if designers know in advance that they can only use two functional evaluations, they will likely prefer to utilize a dichotomous pair to reduce the interval of uncertainty to 0.5 rather than to accept the 0.618 possible through the golden section search. If designers are allowed to adjust the spacing of the merit evaluations as they progress, they may be able to combine the advantages of the symmetrical spacing mentioned previously and the advantages of the dichotomous search to achieve an optimum search algorithm. Suppose that Z_n is the length of the interval of uncertainty after the Nth trial. The symmetry requirement for the evaluations is:

$$Z_{I-2} = Z_{I-1} + Z_I \qquad 1 < I < N,$$

and the requirement that the last evaluation be a dichotomous spacing of ε is:

$$Z_{N-1} = 2 Z_N - \varepsilon.$$

From these two relationships, it is possible to work backwards to determine the required size for any intermediate interval of uncertainty and thus also to determine the placement positions for the merit evaluations. For instance:

$$Z_{N-3} = Z_{N-2} + Z_{N-1}$$
$$= (Z_{N-1} + Z_N) + Z_{N-1}$$
$$= 5Z_N - 2\varepsilon$$

and

$$Z_{N-4} = 8Z_N - 3\varepsilon.$$

The general relationship for any interval of uncertainty is:

$$Z_{N-K} = F_{K+1}(Z_N) - F_{K-1}\varepsilon,$$

where the F_J coefficients are called Fibonacci numbers and are defined in terms of:

$$F_0 = 1$$
$$F_1 = 1$$
$$F_K = F_{K-1} + F_{K-2} \qquad \text{for } K=2,3,\ldots\ldots$$

Table 6-1 presents some of the Fibonacci numbers. In general, the size of the final interval of uncertainty can be expressed as:

$$Z_N = \frac{1}{F_N} + \frac{F_{N-2}}{F_N}(\varepsilon).$$

In the limit, as $\varepsilon \to 0$, a lower bound can be determined on the size of the smallest interval of uncertainty that can be achieved for a given number of evaluations.

Table 6-1 The Fibonacci Numbers

K	F_K
0	1
1	1
2	2
3	3
4	5
5	8
6	13
7	21
8	34
9	55
10	89
11	144
12	233
13	377
14	610
15	987
16	1597
17	2584
18	4181
19	6765
20	10946

The procedure for applying the Fibonacci method is first to decide how many merit evaluations (N) can be used. Then information about the length of the interval of uncertainty can be used to plan the merit evaluation spacings. Since $Z_1 = Z_0 = 1$, this process must start with two evaluations placed Z_2 distance from opposite ends of the initial interval. In this case:

$$Z_2 = \frac{F_{N-1}}{F_N} + \frac{(-1)^N}{F_N} \varepsilon,$$

where ε represents the smallest distance by which two evaluations may be separated and still be distinguished from one another. The next step is to locate an additional evaluation Z_3 units from the end of Z_2. The appropriate interval is retained, and the process is repeated until the Nth evaluation has been carried out.

6.4   A Comparison of One-Dimensional Search Techniques

The five search techniques previously described can best be compared in terms of their efficiency and utility. A common measure of efficiency for an algorithm is the number of functional evaluations required to achieve a desired degree of reduction in the interval of uncertainty. Table 6-2 indicates that the Fibonacci search method is best by this measure and that the total search ranks lowest. Occasionally, the designer is reluctant to utilize the Fibonacci search technique because it requires that the number of function evaluations be specified beforehand. When this situation arises, the golden section search can be used. Because of their efficiency, the designer will usually find that the Fibonacci and the golden section search methods are the best ones to try first on a one-dimensional, unimodal optimization problem.

The utility of an algorithm describes the ease with which it can be applied and the applicability it has to a variety of different problem situations. In terms of utility, the Fibonacci search method is probably the most difficult to apply because it requires a separate calculation to determine the location of the merit evaluation for each new step. This, of course, is the price one must pay to achieve the high efficiency of this algorithm. In terms of utility, the inefficient total search does have one virtue. This is its ability to succeed on nonunimodal merit functions if they are reasonably well behaved. Frequently, the designer does not know whether the merit function is unimodal or not. When this situation exists, the designer should try several different types of algorithms to see if they each converge to the same optimum. This situation points out an important principle in optimization. No single algorithm can solve every

Table 6-2   A Comparison of the One-Dimensional Search Methods

FRACTIONAL REDUCTION OF THE INTERVAL OF UNCERTAINTY†

Number of Merit Evaluations N	Total Search f	Interval Halving f	Dichotomous Search f	Golden Section Search f	Fibonacci Search f
1	1.0	1.0	1.0	1.0	1.0
2	0.667	-	0.500	0.618	0.500
3	0.500	0.500	-	0.382	0.333
4	0.400	-	0.250	0.236	0.200
5	0.333	0.250	-	0.146	0.125
6	0.286	-	0.125	0.090	0.077
7	0.250	0.125	-	0.056	0.048
8	0.222	-	0.0625	0.0345	0.0294
9	0.200	0.0625	-	0.0213	0.0182
10	0.182	-	0.0312	0.0132	0.0112
11	0.167	0.0312	-	0.00813	0.00694
12	0.154	-	0.0156	0.00502	0.00429
13	0.143	0.0156	-	0.00311	0.00265
14	0.133	-	0.00781	0.00192	0.00164
15	0.125	0.00781	-	0.00119	0.00101
16	0.118	-	0.00391	0.000733	0.000626
17	0.111	0.00391	-	0.000453	0.000387
18	0.105	-	0.00195	0.000280	0.000239
19	0.100	0.00195	-	0.000173	0.000148
20	0.095	-	0.000976	0.000107	0.0000913

†All numbers are rounded to three significant digits.

problem. The designer should attempt to become proficient at using a variety of algorithms in order to increase his or her success rate when considering difficult optimization problems.

6.5 Available FORTRAN Codes for One-Dimensional Optimization

In actual practice the engineering designer will probably not write his or her own optimization subroutines for one-dimensional searches. Rather, he or she will make use of software that is presently available. The time and energy saved in utilizing a "ready-made," well-tested FORTRAN algorithm is usually well worth the expense required to obtain it. Table 6-3 presents a compilation of the basic features of available FORTRAN codes suitable for one-dimensional optimization problems. For further details of these, the reader should consult the appropriate reference at the end of this chapter.

Table 6-3   Available FORTRAN Codes for One-Dimensional Optimization

Code Name	Developer/Availability	Method	Special Features or Limitations
COMB1	C. R. Mischke [5] Iowa State University	Total Search	limited to $0.001 \leq f$
GOLD1	C. R. Mischke [5] Iowa State University	Golden Section Search	-
ZXFIB	International Mathematical & Statistical Library [7]	Fibonacci	-

Example 6-2

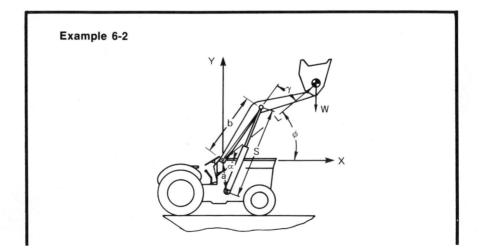

154

Modern loading machines frequently use a slider-crank mechanism with an oscillating slide made of a hydraulic or pneumatic cylinder to implement the deflection of their arm. An example of such a device is shown in the accompanying figure. The design process for this mechanism involves the selection of optimum geometry so that the device performs safely over a desired range of motion. Since the power cylinder is the most expensive element in the lifting portion of this system, optimization for this portion of the design means minimizing the peak load in the cylinder on the range of motion. Clearly, if the load size of the required power cylinder can be reduced, a saving in cost will result.

Recommend a design for the device shown by specifying the attachment points of the hydraulic cylinder if the load to be lifted is W = 1500 kg, the boom length L = 3.0 m, and the motion constraints are:

$$\phi_{min} = -20 \text{ degrees} \qquad S_{min} = 1.0 \text{ m}$$

$$\phi_{max} = 80 \text{ degrees} \qquad S_{max} = 1.8 \text{ m}$$

The problem to be solved is that of finding a mechanism that will satisfy the initial and final positions defined above so that a minimum value is achieved for the maximum cylinder load T on the range of action:

$$\phi_{min} \leq \phi \leq \phi_{max}.$$

(It should be noted that the problem of minimization of the maximum value of a quantity on an interval is a commonly occurring problem in engineering design. This problem is frequently referred to as the "mini-max problem.")

For the mechanism shown in the accompanying figure, the three independent geometric dimensions b, a, and β are the design variables. The relationship between geometry and motion can be found from the law of cosines to be:

$$K_1 \cos(\beta+\phi) - K_2 = -S^2,$$

where:

$$K_1 = 2ab$$

and

$$K_2 = a^2 + b^2.$$

Since the load to be lifted, W, and the boom length, L, will be known prior to the design process, the cylinder load, T, for equilibrium can be found to be:

$$T(\phi) = \frac{LW \sqrt{K_2 - K_1 \cos(\beta+\phi)} \; \cos(\phi)}{ab \sin(\beta+\phi)}.$$

Thus the problem to be solved can be written in standard form as:

The design variables are: a, b, and β.

The merit function to be minimized is:

$$M = [T(\phi)_{max}] \qquad \text{on} \qquad \phi_{min} \leq \phi \leq \phi_{max}.$$

155

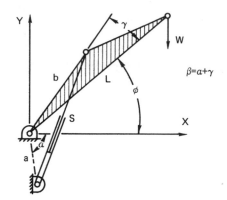

The <u>functional</u> <u>constraints</u> are:

$$K_1 \cos(\beta + \phi_{min}) - K_2 = -S^2_{min}$$

$$K_1 \cos(\beta + \phi_{max}) - K_2 = -S^2_{max}.$$

The need for two of the three design variables can be eliminated through the use of the two equality constraints. This can be accomplished as follows. If the value of β were known, the equality constraints could be used to find:

$$K_1 = \frac{S^2_{min} - S^2_{max}}{\cos(\beta + \phi_{max}) - \cos(\beta + \phi_{min})}$$

and

$$K_2 = K_1 \cos(\beta + \phi_{max}) + S^2_{max}.$$

Because

$$K_1 = 2ab$$

and

$$K_2 = a^2 + b^2,$$

it is possible to determine a and b as follows:

$$b = \left[\frac{K_2 + \sqrt{K_2^2 - K_1^2}}{2}\right]^{0.5}$$

and

$$a = \frac{K_1}{2b}$$

Thus the problem has been reduced to one having only the single design variable β. The value of $T(\phi)_{max}$ may be found by a one-dimensional search scheme on the interval from ϕ_{min} to ϕ_{max}. This will require a separate search apart from the primary optimization search for the best design. Since it is anticipated that the $T(\phi)$ function may be multimodal, it is wise

156

to use a total search scheme for this subsearch even though the overall optimization process may be treated by another method. It should also be noted that it is important to restrict the choices for β so that the value of $T(\phi)$ does not become infinite during the computational process, because this can cause the computer to have difficulty. This situation will happen whenever

$$\sin(\beta+\phi) = 0.$$

To prevent this, it is required that:

$$0 < \beta+\phi < \pi \quad \text{where} \quad \phi_{min} \le \phi \le \phi_{max}.$$

This means that β must be restricted to:

$$-\phi_{max} \le \beta \le \pi - \phi_{min}.$$

A FORTRAN program that solves this problem now follows. This program uses two of the optimization subroutines of Mischke [5] that have been modified to allow the merit function subroutine names to be placed as calling parameters in the argument list of the optimization subroutines. The subroutine COMB performs a total search to find the maximum value of T on the range of motion. This subroutine was selected so that the maximum value of T could be found even if the $T(\phi)$ function were not unimodal. Two of these subroutines are nested for the sake of efficiency in reducing final interval of uncertainty to 0.01 overall. The subroutine GOLD uses the golden section search algorithm to control the overall minimization process.

```
C     *********************************************************
C     *       M A I N    P R O G R A M                        *
C     *  THIS PROGRAM DESIGNS A SINGLE BOOM, LIFT MECHANISM FOR *
C     *  MINIMUM PEAK LOAD IN THE HYDRAULIC CYLINDER.          *
C     *                              T. E. SHOUP 8/30/77       *
C     *********************************************************
C
      EXTERNAL MERIT1,MERIT2
      COMMON S(2),PHI(2),A,B,BETA,W,L,K1,K2
      REAL L,K1,K2
      PI = 3.1415926
C
C     SET GIVEN PARAMETERS
      W = 1500.
      L = 3.0
      S(1) = 1.0
      S(2) = 1.8
      PHI(1) = -PI/9.
      PHI(2) = 4.*PI/9.
C
C     SET THE MAXIMUM AND MINIMUM LIMITS ON BETA
      BTMIN = -1.0*PHI(2)
      BTMAX = PI-PHI(1)
C
C     OPTIMIZE WITH RESPECT TO BETA
      CALL GOLD(1,BTMIN,BTMAX,0.01 ,TMAX,BBEST,B3,B4,J5,MERIT1)
C
C     WRITE THE ANSWERS
      CALL MERIT1(BBEST,TMAX)
      WRITE(6,104)
      WRITE(6,105)A,B,BETA,TMAX
  105 FORMAT(1X,'THE FINAL ANSWERS ARE',/,1X,23('-'),/,
     &       1X,'A   = ',F10.4,'   M',/,
     &       1X,'B   = ',F10.4,'   M',/,
```

157

```fortran
      &         1X,'BETA = ',F10.4,' RAD',/,
      &         1X,'TMAX = ',F10.4,'  KG')
         WRITE(6,104)
  104 FORMAT(1X,23('-'))
C
C     ANALYZE THE MECHANISM FOUND
         WRITE(6,107)
  107 FORMAT(4X,'PHI(RAD)',4X,'LOAD(KG)')
         WRITE(6,104)
         RANGE = (PHI(2)-PHI(1))/50.
         DO 2 I=1,51
         PH = PHI(1) + RANGE*(I-1)
         T = L*W*COS(PH)*SQRT(K2-K1*COS(BETA+PH))/(A*B*SIN(BETA+PH))
    2    WRITE(6,106)PH,T
  106 FORMAT(2(2X,F10.4))
         WRITE(6,104)
         STOP
         END
C     ***********************************************************
C     *     S U B R O U T I N E   M E R I T 1                   *
C     *  THIS SUBROUTINE PRODUCES THE VALUE OF TMAX IF GIVEN    *
C     *  THE MECHANISM PARAMETERS AND A SUITABLE CHOICE FOR     *
C     *  BETA.                                                  *
C     *                             T. E. SHOUP 8/30/77         *
C     ***********************************************************
C
      SUBROUTINE MERIT1(BET,TMAX)
      EXTERNAL MERIT2
      COMMON S(2),PHI(2),A,B,BETA,W,L,K1,K2
      REAL K1,K2
      DIMENSION B3(2000),B4(2000)
      BETA = BET
      K1 = (S(1)**2-S(2)**2)/(COS(BETA+PHI(2))-COS(BETA+PHI(1)))
      K2 = K1*COS(BETA+PHI(2))+S(2)**2
      B = SQRT((K2+SQRT(K2**2-K1**2))/2.)
      A = K1*0.5/B
      CALL COMB(0,PHI(1),PHI(2),0.1,TMAX,B2,B3,B4,B5,B6,J5,MERIT2)
      CALL COMB(0,B5,B6,0.1,TMAX,B2,B3,B4,B5,B6,J5,MERIT2)
C
C     TO MINIMIZE TMAX, USE NEGATIVE MERIT FUNCITON
      TMAX = -1.0*TMAX
      RETURN
      END
C     ***********************************************************
C     *     S U B R O U T I N E   M E R I T 2                   *
C     *  THIS SUBROUTINE PRODUCES THE CYLINDER LOAD FOR AND     *
C     *  GIVEN POSITION OF THE BOOM ARM.                        *
C     *                             T. E. SHOUP 8/30/77         *
C     ***********************************************************
C
      SUBROUTINE MERIT2(PH,T)
      COMMON S(2),PHI(2),A,B,BETA,W,L,K1,K2
      REAL L,K1,K2
      T = L*W*COS(PH)*SQRT(K2-K1*COS(BETA+PH))/(A*B*SIN(BETA+PH))
      RETURN
      END
```

A portion of the output of this computer program now follows.

```
----------------------
THE FINAL ANSWERS ARE
----------------------
A    =     0.5850   M
B    =     1.2691   M
BETA =     1.2231 RAD
TMAX = -7965.8453   KG
----------------------
  PHI(RAD)    LOAD(KG)
----------------------
   -0.3491   7426.0494
   -0.3142   7452.7969
   -0.2793   7482.4390
   -0.2443   7514.2879
   -0.2094   7547.7314
   -0.1745   7582.2217
   -0.1396   7617.2642
   -0.1047   7652.4100
   -0.0698   7687.2479
   -0.0349   7721.3993
   -0.0000   7754.5122
    0.0349   7786.2579
    0.0698   7816.3263
    0.1047   7844.4229
    0.1396   7870.2657
    0.1745   7893.5826
    0.2094   7914.1086
    0.2443   7931.5836
    0.2793   7945.7503
    0.3142   7956.3517
    0.3491   7963.1287
    0.3840   7965.8185
    0.4189   7964.1522
    0.4538   7957.8516
    0.4887   7946.6274
    0.5236   7930.1765
    0.5585   7908.1779
    0.5934   7880.2906
    0.6283   7846.1486
    0.6632   7805.3571
    0.6981   7757.4869
    0.7330   7702.0692
    0.7679   7638.5878
    0.8029   7566.4718
    0.8378   7485.0851
    0.8727   7393.7159
    0.9076   7291.5618
    0.9425   7177.7137
    0.9774   7051.1352
    1.0123   6910.6367
    1.0472   6754.8459
    1.0821   6582.1672
    1.1170   6390.7331
    1.1519   6178.3427
    1.1868   5942.3797
    1.2217   5679.7095
    1.2566   5386.5414
    1.2915   5058.2446
    1.3265   4689.1012
    1.3614   4271.9638
    1.3963   3797.7783
----------------------
```

6.6 Summary

Although the topic of one-dimensional optimization is actually a subset of the more general topic, multivariable optimization, one-dimensional search needs occur so frequently in design, and their methodology is so unique, that they have been discussed separately. In the next chapter the concept of optimization is expanded to the more challenging, multi-dimensional search methods.

Problems

6.1 It is desired to design a new type of storage container consisting of an open, circular cone as shown in the accompanying figure. If the container is to hold a volume of exactly 1.0 cubic meter, what will be the dimensional size of the design having minimum surface area?

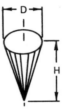

6.2 If the container in Problem 6.1 has a circular lid, find the dimensional size of the design having minimum surface area.

6.3 If the dimension x_1 in Example 6-1 is set to a value of 1.5 m, find the solution to the problem.

6.4

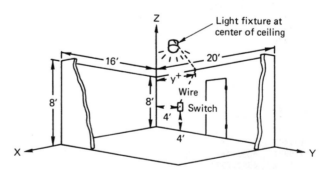

A new, cast concrete, modular concept has been proposed for the construction of hotel rooms. The prefabricated rectangular modules are to be "prewired" before being placed in position. Since a large number of these units will be made to identical specifications, it is essential that the cost of materials be minimized. As part of the design task you have been asked to specify the location of the electrical wire from a wall switch to the ceiling fixture. The design is to use a minimum length of wire and the wire must always be in the wall or in the ceiling. Specify the path that the wire must travel by locating the dimension y*, and specify the total nominal length of wire required for this task.

6.5 A rectangular highway sign of dimensions w and h is to be manufactured from sheet metal of uniform thickness. The sign must have a printed area of 1.5m², a 20 cm margin at the bottom and 10 cm margins along the other three sides as indicated in the accompanying figure. If it is desired to use minimum material, find the best choice for the dimensions w and h.

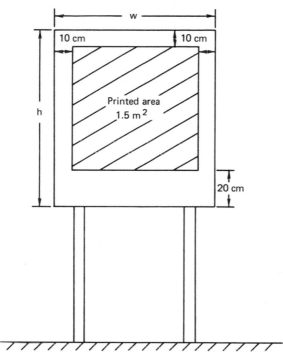

6.6 How would your answer to Problem 6.5 be different if the following constraints were added:

$$1.4 \leq w \leq 2.0 \text{ m?}$$

6.7 Calculate the maximum range (D) and the corresponding firing angle (θ) for the cannon shown in the accompanying figure:

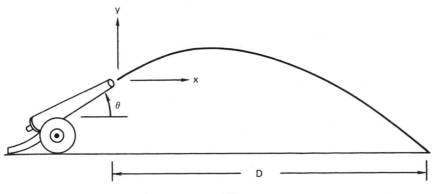

The initial velocity of the projectile is

$$V_o = 800 \text{ m/sec.}$$

The mass of the projectile is:

$$m = 50 \text{ kg.}$$

Assume that the firing takes place on a flat plane $[y(0) = y_{final}]$ and that the drag on the projectile is:

$$F_d = C \, V^2,$$

where $C = 0.015 \text{ N sec}^2/\text{m}$.

6.8 Under what conditions would an engineering optimization problem have no solution? Can you think of an example of this situation?

6.9 The discharge flow rate for steady, fully developed laminar flow of an incompressible fluid through the annular space between two concentric tubes of circular cross section is:

$$Q = K \left[r_o^{\,4} - r_i^{\,4} - \frac{(r_o^{\,2} - r_i^{\,2})}{\ln(r_o/r_i)} \right],$$

where K is a constant that depends on the pressure drop per unit length, the fluid density, and the viscosity. It is desired to design the geometry (choose r_i and r_o) of the two concentric cylinders so as to maximize the flow rate through a cross section area of 10 cm^2 subject to the restriction that:

$$2.0 \leqq r_i < r_o \leqq 10. \text{ cm.}$$

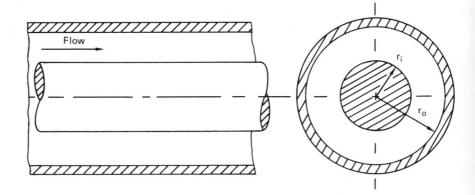

6.10 One creative way to find a root of a general polynomial is by minimizing a merit function that approaches zero magnitude when a root is approached. Try this method using a one-dimensional search method for the polynomial:

$$x^5 + 21x^4 + 158x^3 + 502x^2 + 609x + 245 = 0.$$

References

1. Converse, A. O. Optimization. New York: Holt, Rinehart and Winston, 1970.

2. Fox, R. L. Optimization Methods for Engineering Design. Reading Mass.: Addison-Wesley Publ. Co., 1971.

3. Ketter, R. L., and Prawel, S. P. Modern Methods of Engineering Computation. New York: McGraw-Hill Book Co., 1969.

4. Kuester, J. L., and Mize, J. H. Optimization Techniques with FORTRAN. New York: McGraw-Hill Book Co., 1973.

5. Mischke, C. R. An Introduction to Computer-Aided Design. Englewood Cliffs, N.J.: Prentice-Hall, Inc., 1968.

6. Zahradnik, R. L. Theory and Techniques of Optimization for Practicing Engineers. New York: Barnes & Noble, 1971.

7. International Mathematical and Statistical Library (IMSL). IMSL Library 1, edition 5 (1975) Available trom IMSL, 7500 Bellaire Blvd., Houston, Texas 77036.

7 OPTIMIZATION—PART II

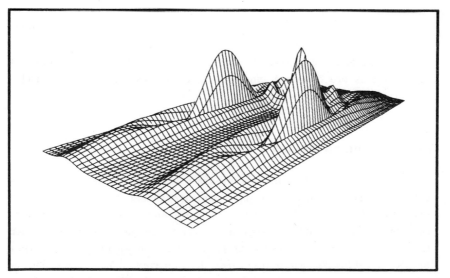

A multimodal merit function with two design variables.
[Computer plot courtesy of California Computer Products, Inc.
(CalComp).]

7.1   Multidimensional Search Techniques

At first thought the designer may be tempted to believe that the difference between multidimensional search techniques and single-dimensional search techniques is only one of increased effort, and that if one is willing to spend a bit more time in the calculation process, he or she could extend single variable methods to N-dimensional methods. Unfortunately, this is not true since the nature of multidimensional space is considerably different from one-dimensional space. For one thing, as the number of dimensions increases, the likelihood that a merit function will be unimodal decreases. In addition, the size of multidimensional space is overwhelming. The amount of effort required to achieve a given degree of reduction of the interval of uncertainty increases by the exponent of the dimensionality of the space. Thus, for instance, if in one-dimensional space, 19 evaluations are needed to achieve $f = 0.1$, then 361 evaluations will be required to achieve this same accuracy in two dimensions, 6,859 in three dimensions, 130,321 in four dimensions and 2,476,099 in five dimensions. Since it is not uncommon to have five or more design variables in a general optimization problem, the seriousness of multidimensionality becomes painfully obvious.

Traditionally, optimization methods in multidimensional space are classified in terms of two broad categories called direct methods and indirect methods. Direct methods utilize a comparison between functional evaluations, while indirect methods employ the mathematical principles of maximization or minimization. Direct methods attempt to establish a strategy to "zero in" on the optimum, while indirect methods attempt to satisfy the conditions of the problem without examining nonoptimal points. In this chapter the basic rationale for the most commonly used, multidimensional optimization algorithms is discussed. Some FORTRAN versions of these methods are compared, and general guidelines for the selection of a specific algorithm are presented.

7.2   Sectioning Method

A logical extension of the one-dimensional search methodology discussed previously would be to alter one independent design variable at a time until the merit value ceases to improve and then to do the same to each individual variable in sequence. Once the last variable has been treated the designer can repeat the process starting with the first variable to see if additional improvement can be achieved. This sectioning or "one-at-a-time" search method will not always reach the optimum. Figure 7-1(a) illustrates a case of contours that are well suited for this tech-

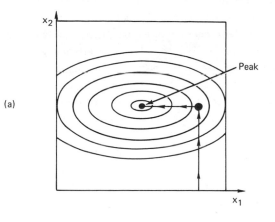

(a)

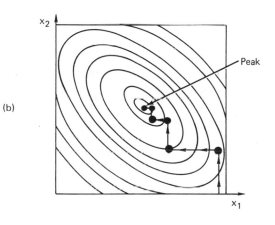

(b)

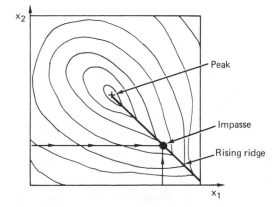

(c)

Figure 7-1 Details of the sectioning method.

nique. The characteristic of this surface is that its contours approximate either circles or ellipses with their major or minor axes parallel to the coordinate axes. If these contour axes are tilted relative to the coordinate axes as shown in Figure 7-1(b), the algorithm efficiency decreases because the method is forced to use many more evaluations to achieve an optimum answer. The method fails completely whenever it encounters a ridge where the contours come to a point, as shown in Figure 7-1(c). Since ridges of this type can frequently occur in engineering design problems, this technique should not be used unless the designer can be sure that the particular problem is free of this pitfall. Nevertheless, the sectioning procedure does find application as a starting procedure for more complex search techniques. The one advantage to this method is that it allows the designer to use one-dimensional search algorithms (such as the golden section search) that are well understood.

7.3   Area Elimination

After seeing how effective one-dimensional techniques are at reducing the interval or area of uncertainty, one might wish to consider the existence of similar methodology in multidimensional space. One of the most obvious area elimination schemes is called the "contour tangent method" because it makes use of a tangent to the contours of the merit function. This technique can best be visualized in terms of the top view of a two-dimensional design merit surface, as shown in Figure 7-2. Here

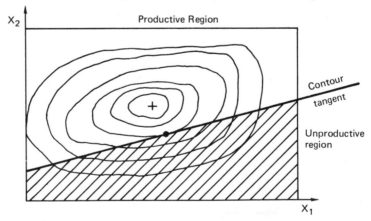

Figure 7-2   The contour tangent method of area elimination.

an arbitrarily selected, feasible point in design space has been found to lie on a contour somewhere below the optimum peak. In the plane of the contour, the tangent to the contour can be located. This tangent is not

168

difficult to find, because it must lie in the plane of the contour and must also be perpendicular to the local gradient of the merit surface at the design point. If the merit surface is well behaved and is strongly unimodal, the contour tangent will divide the feasible design space into regions that have high promise or low promise in so far as the extremum is concerned. Using this technique and several well placed merit observations, the designer can substantially reduce the search domain. There is a difficulty associated with the use of this algorithm, however. If a contour line is concave rather than convex, as shown in Figure 7-3, the

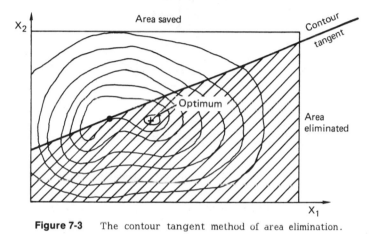

Figure 7-3   The contour tangent method of area elimination.

possiblity of eliminating the region containing the extremum does exist. In addition, the region of uncertainty remaining after several eliminations may be awkward to manipulate by other algorithms.

One type of area elimination technique that does give manageable results is the grid search of Mischke [15]. In this technique the reduced area of uncertainty will be a hypercube (i.e., the multidimensionalization of a square or cube) of predictable size. For this reason it is one of the few multidimensional techniques with measurable effectiveness. Visualization in terms of a design space with two design variables will aid in the understanding of this search technique. The original region of uncertainty is mapped onto a unit square, cube, or hypercube (depending on the dimension of the space) so that the search is normalized to a region of unit dimension on a side. Through this hypercube is drawn a grid of symmetrically paired, orthogonal planes parallel to the design variable axes. The intersections of planes will generate lines that will in turn intersect to give points known as nodes, as shown in Figure 7-4. The merit values at each intersection and at the center of the cube are evaluated. This involves $2^M + 1$ total functional evaluations for "M" design

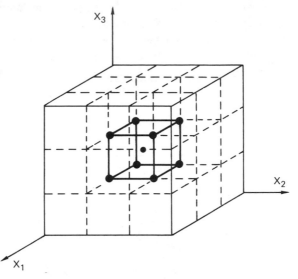

Figure 7-4   The grid search technique.

variables. A note is made of the largest merit value found, and this location becomes the center of a smaller hypercube for further investigation. The process continues until the desired degree of reduction in the interval of uncertainty is achieved. If the fractional reduction in allowable domain along any design variable axis is "r" then the linear fractional reduction for "b" hypercubes will be:

$$f = r^b$$

and the total function evaluations will be:

$$N = b(2^M) + 1.$$

Mischke [15] recommends that

$$2/3 < r < 1$$

be used and comments that a star pattern is more efficient than grid node points for problems with design variables of three or greater.

7.4   Random Search

Earlier in this chapter, the vastness of multidimensional space was discussed in terms of how many evaluations would be necessary to achieve $f = 0.1$ by a lattice approach. It was found that the number increases by the exponent of the dimension of the space. An interesting and creative alternative to circumvent this difficulty has been proposed by Brooks[1] and is based on a random selection process. Suppose that the design space is viewed as a unit cube or hypercube and is divided into cubical cells by means of 10 equally spaced divisions along each design variable axis. If N = 2, there will be 100 cells; if N = 3 there will be 1000 cells; and, in general, there will be 10^N cells for a space having "N" design variables. The probability that one cell chosen at random will be among the best 10% is 0.1, because we would be looking for the one best in 10 for N = 1, one of the ten best in 100 for N = 2, etc. The probability of missing one of the best 10 percent will be 0.9. If two cells are selected at random, the probability of missing becomes $(0.9)^2$ or 0.81. In general, the probability of finding at least one cell in the best fraction f, when N random selections are made is:

$$P = 1 - (1-f)^N.$$

Table 7-1 provides a listing of the number of random selections needed to achieve a specific probability for a desired best fraction. From this table it can be seen that 44 random selections will have a 99% probability of achieving f = 0.1. This represents an attractive alternative to the 2,476,099 evaluations required to guarantee f = 0.1 by total enumeration with five design variables.

Table 7-1

f	Probability			
	0.80	0.90	0.95	0.99
0.1	16	22	29	44
0.05	32	45	59	90
0.01	161	230	299	459
0.005	322	460	598	919

The random process has two attractive features. First, the method works well on any type of merit surface whether unimodal or not. Second, the probability of success in N random selections does not depend on the dimension of the space being considered. Although this technique does not move toward an optimum, it does provide a way to achieve a "good" starting point to be used by other search methods. For this reason it is often used in combination with one or more algorithms of a different type.

7.5   Gradient Methods

A large number of multidimensional optimization algorithms depend in some way on gradient information. The basis for this fact can be seen in a simple illustration. Suppose that a mountain climber were blindfolded and told to climb to the top of a unimodal mountain. Even without the benefit of being able to see the peak, the climber could reach the top simply by remembering always to walk uphill. Although any rising path will eventually lead to the top, the path where the slope is steepest is the best, provided that the climber does not encounter a vertical cliff that must be scaled. (The mathematical equivalent of a cliff would be a ridge caused by a constraint in the merit surface.) For now it will be assumed that the optimization problem is unconstrained. Later the concept of constraints will be incorporated into the search scheme. The optimization equivalent of the steepest path idea is known as the method of steepest ascent or the method of steepest descent. The gradient vector is perpendicular to a contour and can be used to locate a new design point. It should be noted that the gradient method, unlike the contour tangent method, can be used on any unimodal function, not just those that are strongly unimodal.

In order to understand the rationale of gradient based methods it is well to look into the nature of the gradient. Consider a system of independent unit vectors $\vec{e}_1$, $\vec{e}_2$, $\vec{e}_3$, $...,\vec{e}_N$ that are aligned with the design variable axes x_1, x_2, x_3, $...x_N$. The gradient vector for a general merit function $F(x_1, x_2, x_3, ...,x_N)$ is of the form:

$$\text{gradient} = \frac{\partial F}{\partial x_1}\,\vec{e}_1 + \frac{\partial F}{\partial x_2}\,\vec{e}_2 + ..., + \frac{\partial F}{\partial x_N}\,\vec{e}_N,$$

where the partial derivatives are evaluated at the point being considered. This vector points in the upward or ascent direction, and its negative points in the descent direction. The unit gradient vector is often written as:

$$\text{gradient} = v_1\vec{e}_1 + v_2\vec{e}_2 + v_3\vec{e}_3 + .., + v_N\vec{e}_N,$$

where:

$$v_i = \frac{\dfrac{\partial F}{\partial x_i}}{\left[\displaystyle\sum_{j=1}^{N}\left[\left(\frac{\partial F}{\partial x_j}\right)^2\right]\right]^{\frac{1}{2}}}$$

In some cases the nature of the merit function is well known enough to allow direct differentiation to calculate the gradient vector components. If

172

the partial derivatives cannot be extracted in this way, they may be approximated by small local explorations to get:

$$\frac{\partial F}{\partial x_i} = \frac{F(x_1, x_2, ..,x_i+\Delta, ..,x_N) - F(x_1, x_2, ..,x_i, ...,x_N)}{\Delta},$$

where Δ is a small exploration along the x_i direction. This formula is often referred to as the "secant approximation." Once the gradient direction is known, it can be used in a variety of ways to implement a search strategy.

Steepest Ascent by Steps

Some search methods move a fixed step size up the gradient and reevaluate the function. If an improvement has been achieved, a new gradient is computed and the procedure is repeated, often with an increased step size. If no improvement or a negative improvement is found, the step size from the previous best point is decreased and the procedure is repeated. This process continues until no improvement can be achieved by decreasing the step size.

Steepest Ascent by One-Dimensional Search

Some search methods use information about the gradient to conduct a one-dimensional search along the direction of the steepest ascent or descent using the relationship:

$$x_i(new) = x_i(old) + Sv_i,$$

where S is the new one-dimensional parameter along the gradient. Once the one-dimensional optimum along the gradient has been achieved, a new gradient is found, and the process is repeated until no further improvement can be found. The primary advantage to this method is that the parameter S may be used as the independent variable for a Fibonacci search, and thus the method tends to be quite efficient in effort. One of the prime advantages of the steepest gradient methods is their ability to steer away from saddle points on the merit surface, as indicated in Figure 7-5. It should be noted in this figure, however, that gradient techniques will find only a local optimum when applied to multimodal surfaces. For this reason, if the nature of the surface is not well known, several starting points should be tried to see if every start leads to the same optimum. Another difficulty that can hamper the efficiency of the gradient method occurs when the technique encounters a ridge. Because a ridge represents a discontinuity in the slope of a contour line, it tends to give false information about the proper direction to move. Thus the search technique may slow down and "zig-zag" back and forth across the ridge,

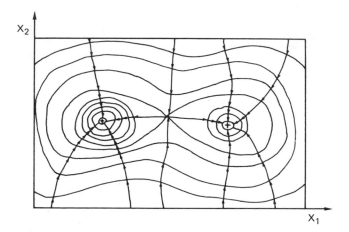

Figure 7-5 A bimodal merit surface.

making progress so slow that the algorithm often must be abandoned. In reality, however, most response surfaces have one or more ridges. These ridges often point toward the optimum. Thus, whenever a ridge is encountered, the best direction to move is along the ridge. Several sophisticated search techniques have been designed to exploit this ridge behavior. These techniques are referred to as "accelerated climbing techniques." Several of these are discussed next along with other "non-gradient" types of algorithms that work well on merit surfaces with ridges.

7.6 Fletcher-Reeves Method

The Fletcher-Reeves [10] method is an optimization algorithm for finding the unconstrained minimum of a multivariable nonlinear merit function of the form:

$$\text{Merit} = F(x_1, x_2, \ldots, x_N).$$

The method uses derivatives of the merit function with respect to the independendent variables. Unimodality is assumed, and, therefore, when using this method, the designer should try multiple starting points if multimodality is suspected. The logic diagram for this method is shown in Figure 7-6. The algorithm proceeds as follows. First, a feasible starting point in the design space is selected, and the direction of steepest descent is calculated in terms of the vector components:

174

$$v_i^{(k)} = \frac{-\partial F/\partial x_i^{(k)}}{\left[\displaystyle\sum_{j=1}^{N}\left(\frac{\partial F}{\partial x_j}\right)^2\right]^{\frac{1}{2}}} \qquad i = 1, 2, \ldots, N,$$

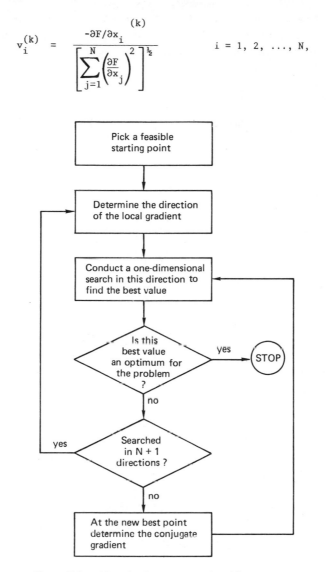

Figure 7-6 The Fletcher-Reeves algorithm.

where $k = 1$ for the starting point. Next, a one-dimensional search is conducted along the direction of steepest descent using the relationship:

$$x_i(\text{new}) = x_i(\text{old}) + Sv_i \qquad i = 1, 2, \ldots, N,$$

where S represents the distance moved along the gradient vector. When a minimum is reached in this one-dimensional search, new, unit search directions are computed. These new directions are not exactly along the new gradient vector but are composed of a linear combination of the present gradient and the previous gradient. The new, unit direction vector components are found from:

$$v_i^{(k+1)} = \frac{-\dfrac{\partial F}{\partial x_i}^{(k+1)} + \beta^{(k)} v_i^{(k)}}{\left\{ \displaystyle\sum_{j=1}^{N} \left[-\left(\frac{\partial F}{\partial x_j}\right)^{(k+1)} + \beta^{(k)} v_j^{(k)} \right]^2 \right\}^{\frac{1}{2}}}, \quad i = 1, 2, \ldots, N$$

where:

$$\beta^{(k)} = \frac{\displaystyle\sum_{i=1}^{N} \left[\left(\frac{\partial F}{\partial x_i}\right)^{(k+1)} \right]^2}{\displaystyle\sum_{i=1}^{N} \left[\left(\frac{\partial F}{\partial x_i}\right)^{(k)} \right]^2}$$

The iteration index "k" denotes the sequence of the calculation in the iteration process. The new directions are said to be "conjugate" directions and correspond to the current local quadratic approximation to the function. A one-dimensional search is then conducted along this new direction; when a minimum is found, a convergence check is made. If the check reveals that convergence is achieved, the procedure stops. If convergence is not achieved, a new set of conjugate directions is calculated, "k" is indexed by one, and the process is repeated until convergence is achieved or until N + 1 conjugate direction steps have been searched. After a cycle of N + 1 directions have been used, a new cycle is started using the steepest descent direction once again. The rationale for this algorithm is based on the prospect of exploiting the propensity that gradient techniques have for finding ridges in the design space. Because the second through N + 1 search directions are not along the gradient, they do not tend to "hang up" on the ridge but rather tend to exploit the notion that a ridge will usually point toward the extremum. It is true, in general, that methods that calculate a new search direction based on accumulated knowledge of the local behavior of the function are inherently more powerful than those in which the directions are assigned in advance. For this reason the Fletcher-Reeves method is far superior to steepest ascent or descent methods. The disadvantage to the method is that, because it is more complex than steepest gradient methods, it usually requires a correspondingly more complex subroutine for implementation.

176

Example 7-1

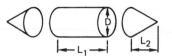

It is desired to design a new type of aircraft fuel storage pod as shown in the accompanying figure. The symmetrical pod is to be fabricated from three sections of metal sheet material and will consist of a cylindrical center with conical ends. The three pieces, once fabricated, will be welded together. The design specifications for this pod call for a volume of 1.0 m³. It is desired to minimize the amount of material used in the design for reasons of weight and cost.

What dimensions L_1, L_2 and D would you recommend for this design?

For this design problem, the quantity to be minimized will be the total surface area of the pod:

$$A = L_1 \pi D + 2[\pi(D/2) \sqrt{(D^2/4)+L_2^2}].$$

The need for one of the three design variables L_1, L_2, and D can be eliminated using the equality constraint:

$$\text{Volume} = 1.0 = \pi D^2[(L_2/6)+(L_1/4)].$$

Thus:

$$D = \sqrt{\frac{1}{\pi[(L_2/6)+(L_1/4)]}}$$

Thus the problem to be solved can be written in standard form as:

The <u>design variables</u> are: L_1 and L_2.

The <u>merit function</u> to be minimized is:

$$A = \sqrt{\frac{\pi}{[(L_2/6)+(L_1/4)]}} \left[L_1 + \sqrt{\frac{1}{4\pi[(L_2/6)+(L_1/4)]}} + L_2^2\right].$$

Since this problem is not constrained, it is likely that the optimum design will lie at a point on the merit surface where the local gradient is zero. Since taking partial derivatives of the merit function and solving for this point can be cumbersome, and since no ridges (due to constraints) are anticipated in the merit surface, a gradient based search scheme will be employed. The method to be used is the Fletcher Reeves algorithm [10]. A FORTRAN program that implements this algorithm now follows. This program makes use of the FMCG subroutine of the IBM Scientific Subroutine Package [13]. The merit function subroutine FUNCT that is required by subroutine FMCG must provide not only the merit ordinate at a design point but also the local gradient. In order to avoid the need for partial derivatives, subroutine FUNCT approximates the gradient by making small incremental changes in the design values and noting the incremental change in merit.

Since it is possible that the unconstrained search algorithm may undertake search excursions into areas where one or both of the design variables have negative values, absolute values of the design variables are used in the merit function subroutine to ensure that these negative values do not appear to give a more favorable surface area. The subroutine FMCG requires that the user provide an initial estimate for the optimum merit value and an initial selection of design values to specify where the search should begin. Since a sphere has the most volume per surface area of any closed container, the surface area of a sphere having a volume of 1 cubic meter will provide a lower bound (A = 4.84) for the surface area of the fuel pod in this problem. Thus the value EST = 5.0 is selected for the estimate of the optimum surface area. The initial selection of design variables (L_1 = 1.0 and L_2 = 0.5) is arbitrary.

```
C     **********************************************************
C     *           M A I N    P R O G R A M                     *
C     *  THIS PROGRAM PERFORMS THE DESIGN OF A SYMMETRICAL      *
C     *  CYLINDRIAL TANK WITH CONICAL ENDS.  THE TANK           *
C     *  MUST HAVE A VOLUME OF 1 CUBIC METER AND MINIMUM        *
C     *  SURFACE AREA.                                          *
C     *                             T. E. SHOUP   9/8/77        *
C     **********************************************************
C
      EXTERNAL FUNCT
      REAL L(2)
      DIMENSION G(2), H(4)
      PI = 3.1415926
C
C     AN ESTIMATE OF THE MINIMUM AREA IS
      EST = 5.0
C
C     A TEST VALUE REPRESENTING THE EXPECTED ABSOLUTE ERROR
      EPS = 0.000001
C
C      SET STARTING VALUES
      L(1) = 1.
      L(2) = 0.5
C
C     MINIMIZE THE AREA BY THE METHOD OF FLETCHER REEVES
C
      CALL FMCG (FUNCT, 2, L, F, G, EST, EPS, 400, IER, H)
C
C     COMPUTE THE DIAMETER
      D = SQRT(1./(PI*(L(2)/6. + L(1)/4.)))
C
C     WRITE THE ANSWERS
      WRITE (6,100) IER, F, L(1), L(2), D
  100 FORMAT (1X,23('-'),/,1X,'THE FINAL ANSWERS ARE',/,1X,
     &        'IER =',I10,/,1X,'AREA =',F10.5,' M**2',/,1X,
     &        'L(1) =',F10.5,'    M',/,1X,'L(2) =',F10.5,'    M',
     &        /,1X, 'D    =',F10.5,'    M',/,1X,23('-'))
      STOP
      END
C     **********************************************************
C     *           S U B R O U T I N E    F U N C T              *
C     *  THIS SUBROUTINE PRODUCES THE TANK SURFACE AREA         *
C     *  AND AREA GRADIENT GIVEN THE TANK DIMENSIONS.           *
C     *  THE GRADIENT IS APPROXIMATED BY INCREMENTAL            *
C     *  CHANGES.                                               *
C     *                             T. E. SHOUP   9/8/77        *
C     **********************************************************
      SUBROUTINE FUNCT (N, L, AREA, GRAD)
      REAL L(2), GRAD (2), L1, L2
```

```
      A(L1, L2, PI) = SQRT (PI/(ABS(L2)/6. + ABS(L1)/4.))*
     &    (ABS(L1) + SQRT(1./(4. * PI * (ABS(L2)/6. + ABS(L1)/4.))
     &+ L2 **2))
C
      PI = 3.1415926
C
      AREA = A(L(1), L(2), PI)
C
C     COMPUTE GRADIENTS
      X1L = L(1)*0.999
      X1R = L(1)*1.001
      GRAD(1) = (A(X1L,L(2),PI)-A(X1R,L(2),PI))/(X1L-X1R)
      X2L = L(2)*0.999
      X2R = L(2)*1.001
      GRAD(2) = (A(L(1),X2L,PI) - A(L(1),X2R,PI))/(X2L-X2R)
C
      RETURN
      END
```

The output of this computer program is:

```
------------------------
THE FINAL ANSWERS ARE
IER =        0
AREA =    5.01921  M**2
L(1) =    0.53479   M
L(2) =    0.53454   M
D    =    1.19531   M
------------------------
```

The value IER=0 means that convergence was obtained.

7.7   Davidon-Fletcher-Powell Method

The Davidon-Fletcher-Powell [9] method is an optimization algorithm for finding the unconstrained minimum of a multivariable merit function of the form:

$$\text{Merit} = F(x_1, x_2, \ldots, x_N).$$

Derivatives of the merit function with respect to the independent variables are necessary. Since the algorithm is based on the assumption of uni-modality, several alternate starting points are recommended if the merit surface is suspected to be multimodal. The logic diagram for this method is shown in Figure 7-7. The algorithm can be described as follows. First, a feasible point in design space is selected as a starting location. The direction of search is computed using the vector components:

$$v_i^{(k)} = \frac{\displaystyle\sum_{j=1}^{N} H_{i,j} \frac{\partial F}{\partial x_j}^{(k)}}{\left\{ \displaystyle\sum_{\ell=1}^{N} \left[\sum_{j=1}^{N} H_{\ell,j} \frac{\partial F}{\partial x_j} \right]^2 \right\}^{\frac{1}{2}}}, \qquad i = 1, 2, \ldots, N,$$

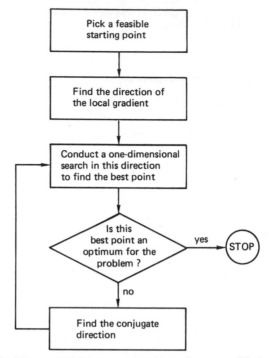

Figure 7-7 The Davidon-Fletcher-Powell algorithm.

where k represents the iteration index and $\underline{H}_{i,j}$ represents the elements of a symmetrical, positive-definite N × N matrix. As the iteration process proceeds, this matrix becomes the inverse of the Hessian matrix (the matrix of second partial derivatives) of the merit function. Since this matrix is generally not known in advance, any symmetrical, positive-definite matrix can be used to start the process. The usual choice is to use the identity matrix. When this particular selection is made, the initial direction of search is along the line of steepest descent. A one-dimensional search is conducted along this initial search direction using the relation:

$$x_i(\text{new}) = x_i(\text{old}) + S v_i \qquad i = 1, 2, \ldots, N,$$

where S is the step size in the direction of the search. A convergence check is made once the one-dimensional optimum is reached. If convergence is achieved, the search is terminated. If convergence is not achieved, a new search direction is chosen using the previous relationship and a new H matrix found as follows:

$$\underline{H}^{(k+1)} = \underline{H}^{(k)} + \underline{A}^{(k)} - \underline{B}^{(k)}.$$

The elements of the $N \times N$, $\underline{A}^{(k)}$ and $\underline{B}^{(k)}$ matrices are computed by means of the following formulas (the superscript "t" denotes "transpose"):

$$\underline{A}^{(k)} = \frac{\Delta \underline{x}^{(k)} (\Delta \underline{x}^{(k)})^t}{(\Delta \underline{x}^{(k)})^t (\Delta \underline{G}^{(k)})}$$

$$\underline{B}^{(k)} = \frac{\underline{H}^{(k)} \Delta \underline{G}^{(k)} (\Delta \underline{G}^{(k)})^t (\underline{H}^{(k)})}{(\Delta \underline{G}^{(k)})^t \underline{H}^{(k)} \Delta \underline{G}^{(k)}} .$$

In these expressions, $\Delta \underline{x}^{(k)}$ and $\Delta \underline{G}^{(k)}$ represent column vectors of differences in the x_i values and differences in the gradient values between locations. These column vectors are:

$$\Delta \underline{x}^{(k)} = \underline{x}^{(k+1)} - \underline{x}^{(k)} \qquad \text{(difference in location between iterations)}$$

$$\Delta \underline{G}^{(k)} = \frac{\partial F}{\partial \underline{x}}^{(k+1)} - \frac{\partial F}{\partial \underline{x}}^{(k)} \qquad \text{(difference in gradients between iterations)}$$

Owing to the nature of the matrix operations, in the expressions for $\underline{A}^{(k)}$ and $\underline{B}^{(k)}$, the numerators are each $N \times N$ matrices, and the denominators are scalars. Once the new search directions are known, a new one-dimensional search is performed, and the iterative process continues. By this algorithm the searches, after the first, are in directions of locally improving values of the merit function but are rarely along the gradient. For this reason the algorithm is often called a "deflected gradient" method. Because of this special characteristic, the method is able to negotiate ridges in the design space. Most authorities regard this technique as the most powerful of the gradient-based searches. Unlike the Fletcher-Reeves method, the Davidon-Fletcher-Powell method yields full information on the curvature of the merit function at the minimum. This information is obtained at the price of providing storage space and manipulation time for the matrix $\underline{H}$.

7.8   Hooke and Jeeves Pattern Search

The pattern search technique of Hooke and Jeeves [12] is an easily programmed, climbing technique that does not require the use of derivatives. The algorithm has ridge-following properties and is based on the premise that any set of design moves that have been successful during early experiments will be worth trying again. The method is based on the assumption of unimodality and is used to find the minimum of a multi-variable, unconstrained function of the form:

$$\text{Merit} = F(x_1, x_2, \ldots, x_N).$$

The logic diagram for this method is shown in Figure 7-8. The algorithm

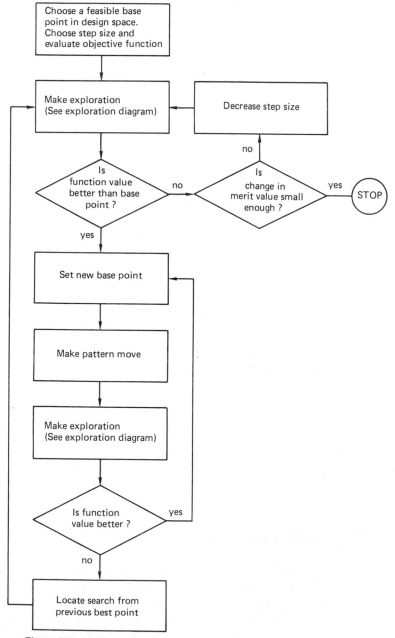

Figure 7-8   The Hooke and Jeeves pattern search algorithm.

proceeds as follows. First, a base point in the feasible design space is chosen along with exploration step sizes. Next, an exploration is performed a given increment along each of the independent variable directions following the logic shown in Figure 7-9. Whenever a functional improvement is obtained, a new temporary base point is established. Once this

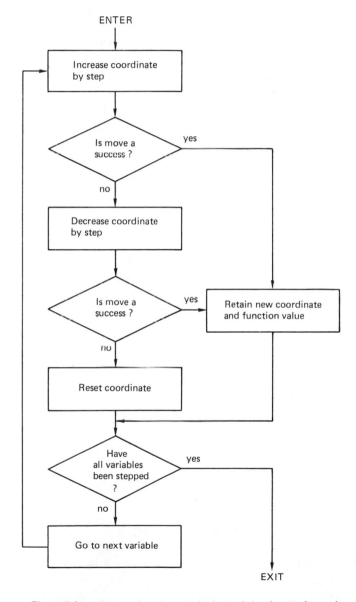

Figure 7-9   The exploration method used in the Hooke and Jeeves algorithm.

exploration is complete, a new base point is established, and a "pattern move" takes place. This pattern move consists of an extrapolation along a line between the new base point and the previous base point. The distance moved beyond the best base point is somewhat larger than the distance between the two base points. Mathematically, this extrapolation is:

$$x_{i,o}^{(k+1)} = x_i^{(k+1} + \alpha(x_i^{(k+1)} - x_i^{(k)}),$$

where $x_{i,o}^{(k+1)}$ becomes a new temporary base point or "head." In this expression, "i" is the variable index, "k" is the stage index, and "α" is an acceleration factor that is greater than or equal to 1.0. Once the new temporary base point has been found, an exploration about this point is instituted to see if a better base point can be found. This exploration also uses the logic of Figure 7-9. If the temporary head or any of its neighboring points are a better base, the pattern process repeats using this improved base. Because of the nature of the acceleration factor, each successive pattern extrapolation becomes bolder and bolder until the process oversteps the peak or a ridge. At this point the previous "best base" is recalled, the local exploration step size is decreased, and the pattern-building process begins again. Once the step size is decreased below a predetermined value and still no substantial change in the merit value can be achieved, the procedure terminates.

For this method, after a few modifications of direction, the pattern coincides with the shape of the ridge. Typically, once a pattern search is established, the pattern move may grow until its length reaches 10 to 100 times the basic step size. Thus, when a pattern move fails, the only way to continue the search is to begin again from the previous best base point with an entirely new pattern. The fact that this algorithm can "accelerate" contributes to its overall efficiency. The method also has the advantage of providing an approximate solution, improving all the while, at all stages of the calculation. Pattern search methodology has proved particularly successful in locating extrema on hypersurfaces that have sharp valleys. On such surfaces gradient techniques behave badly.

7.9   Rosenbrock Pattern Search

The pattern search of Rosenbrock [17] is a ridge-following method that has been shown to be effective often when other methods fail. This method is frequently called the "method of rotating coordinates" because of the way it performs the local explorations. Rather than perturbing each of the original variables independently, this method rotates the coordinate system so that one axis points along a ridge. The location of this ridge is determined by a previous trail. The remaining axes are arranged ortho-gonal to this first axis. The method is based on the assumption of uni-modality and is used to find the minimum of a multivariable, unconstrained function of the form:

$$\text{Merit} = F(x_1, x_2, \ldots, x_N).$$

The logic diagram for this method is shown in Figure 7-10. The algorithm

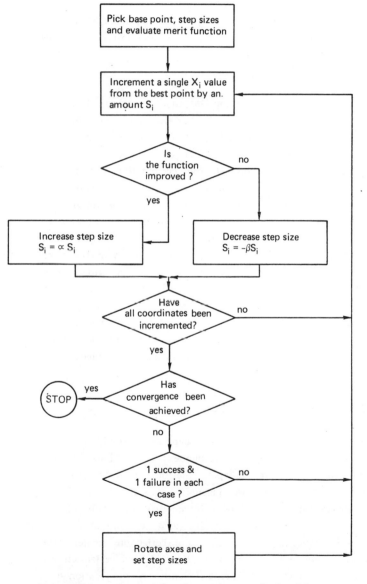

Figure 7-10   The Rosenbrock pattern search algorithm.

proceeds as follows. First a starting point and initial step sizes (S_i, i = 1, 2, ...,N) are picked, and the merit function is evaluated. Then, in turn, each variable x_i is stepped a distance S_i parallel to the design variable axis, and the merit function is evaluated. As this process proceeds, if the value of F decreases, the move is termed a success and the step distance is increased by the formula:

$$S_i = S_i \alpha,$$

where α is a preselected value greater than 1. On the other hand, if the value of F increases, the move is termed a failure, and the step distance is decreased by the formula:

$$S_i = -S_i \beta,$$

where β is a preselected factor less than 1. Once all of the variables have been stepped, a convergence check is made. If the process has converged, the procedure terminates. If the process has not converged, an additional check is made to see if at least one success and one failure have occurred in each direction. If this combination of success and failure has not been achieved, the stepping procedure is repeated starting with the first variable. If at least one success and one failure in each direction have occurred, the axes are rotated so that the initial search direction is in a previously established direction of greatest improvement. Step sizes are then set, and the search continues in each of the variable directions using the new coordinate axes. Instead of moving a fixed step in each direction, this algorithm attempts to find the optimum point on each line. In so doing, the procedure continuously adjusts the step size for the pattern search. The combination of rotation of the ridge-following vector and the adjustment of the scale has made this algorithm an extremely powerful one for handling difficult optimization problems.

7.10 The Simplicial Method

In order to understand the simplicial method, it is first necessary to understand the concept of a simplex. A simplex is an N-dimensional, closed geometric figure in space that has straight line edges intersecting at N + 1 vertices. In two dimensions this figure would be a triangle. In three dimensions it would be a tetrahedron. Search schemes based on the simplex utilize observations of the merit function at each of the vertices. The basic move in this method is a reflection to generate a new vertex point and thus a new simplex. The choice of reflection direction and the choice of the new vertex point depends on the location of the worst point in the simplex, as demonstrated in Figure 7-11. The new point is called

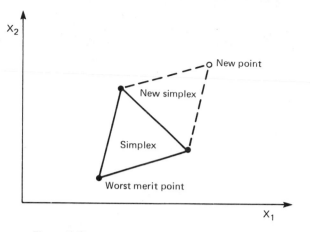

Figure 7-11 A simplex in two-dimensional space.

the "complement" of the worst point. If the newest point in a new simplex has the worst value in that new simplex, the algorithm would oscillate back and forth rather than moving onward to the extremum. When this happens the basic move pattern is modified by using the second worst point to locate the complement. The simplex technique tends to move the centroid of the area it bounds toward the extremum.

If, in addition to the reflection process, the edges of the simplex are allowed to contract and expand in size, the algorithm becomes that presented by Nelder and Mead [16]. This general algorithm adapts to local landscape in order to reach the minimum of a unimodal function of the form:

$$\text{Merit} = F(x_1, x_2, \ldots, x_N).$$

The logic diagram for this method is shown in Figure 7-12. The algorithm can be described as follows. First, the initial simplex is placed in the design space. Then, the vertex points having the worst and best merit values, P (worst) and P (best), are located. Next, the centroid $\bar{P}$ of all points in the simplex, excluding the worst point, is calculated. A reflected point P* is calculated from the expression:

$$P^* = (1 + \alpha)\bar{P} - \alpha\, P(\text{worst}),$$

where α is a positive constant called the reflection coefficient. If this reflection point represents the best merit selection so far, a further expansion is made by the formula:

$$P^{**} = \gamma P^* + (1 - \gamma)\, \bar{P},$$

187

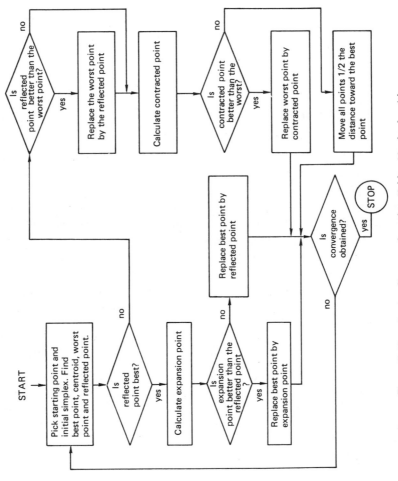

Figure 7-12 Logic diagram for the Nelder-Mead algorithm.

where γ is a preselected constant called the expansion coefficient and is greater than unity. If the merit value of P** is a further improvement, the current value of P(best) is replaced by P**, and the process is restarted after a convergence check. On the other hand, if P** is not better than P*, then P(best) becomes P*, and the process continues.

If P*, as it was previously found, is not better than P(best), then P(worst) is defined to be either the old P(worst) or the value of P*, whichever has the more favorable merit value. Once this decision is made, a contraction point

$$P** = \beta P(worst) + (1 - \beta)\bar{P}$$

is calculated. Here β is a contraction coefficient between 0 and 1. If the contracted point is more favorable than P(worst), the value of this P** replaces P(worst). If this contracted point is not more favorable than the previous P(worst), then all P_i values are replaced by

$$P_i(new) = \frac{P_i(old) + P(best)}{2}$$

before the iteration process is continued.

This algorithm has the versatility to adapt itself to the local landscape of the merit surface. It will elongate down inclined planes; it will change direction on encountering a valley at an angle; and it will contract in the neighborhood of an extremum. The method is quite effective and is computationally compact.

7.11   Penalty Function Methods

Nearly all of the optimization techniques that have been described up to now in this chapter can be categorized as methods for unconstrained minimization or maximization. Many of these methods are quite powerful and offer great potential for use in the design process. Yet it seems clear that many engineering problems have constraints of the form:

$$G_j(x_i) \geqq 0 \qquad j = 1,2,..,J.$$

Therefore, it seems reasonable to look for a way to adapt constrained problems for solution by unconstrained methods. The mathematical device for accomplishing this task is the "penalty function."

The basis for the penalty function method is a new merit function of the form:

$$M(x_i) = F(x_i) + \sum_{j=1}^{N} \phi[G_j(x_i)].$$

In this expression:

$F(x_i)$ is the objective function (constrained),

$M(x_i)$ is the composite function (unconstrained), and

$\phi(G_j(x_i))$ is a penalty function based on the inequality constraints. The composite merit function is formed by summing the penalty function and the previous objective function. As an example of a penalty function, consider a function $\phi(x_i)$, which is zero for all design points satisfying the constraints $G_j(x_i) \geq 0$ and is infinite for all design points that violate the constraints. Clearly, if all constraints are satisfied, the minimization of $M(x_i)$ is equivalent to the minimization of $F(x_i)$. If, on the other hand, any constraint is violated, the merit function goes to infinity, which is far from the minimum of $F(x_i)$. Thus any design that violates the constraints is said to be "penalized."

There do exist certain difficulties associated with the implementation of the penalty function suggested previously. To overcome many of these, a few researchers [2,3,6,7,8] have shown that a gradual sequence of penalty function minimizations is desirable. Thus, instead of solving only one unconstrained optimization problem, a sequence of problems is considered, each of which comes closer to the final desired solution. The Fiacco-McCormick [6,7,8] method suggests the use of a penalty function of the form:

$$\phi[G_j(x_i)] = r_p G_j^{-1}(x_i) \qquad p = 1,2,\ldots,$$

where "r_p" is the penalty parameter, and the index "p" identifies the successive values of the penalty parameter for the sequence of problems considered. Thus the unconstrained merit function becomes:

$$M(r_p, x_i) = F(x_i) + r_p \sum_{j=1}^{J} \frac{1}{G_j(x_i)} \qquad p = 1,2,\ldots,$$

Because of the nature of the inverse function, the merit value M will approach infinity on the constraint boundary. The solution method using this penalty function proceeds as follows. First, a feasible design point in the allowable space is chosen inside the constrained set. Using any desired technique, the optimum is found for a preselected starting value of r_1. Because the starting point is inside of the constrained area, the optimum found will also be within this area. Any trajectory of steepest descent leading from the internal design point cannot penetrate the boundary. Once the minimum is achieved, its location becomes the starting point for a new merit function in which the value of r_p is reduced. The optimization process is repeated for a sequence of decreasing r_p values:

$$r_1 > r_2 > r_3 > r_4 \ldots > 0.$$

As r_p approaches zero, the solution to the unconstrained optimization problem approaches the solution to the constrained problem.

Two simple "penalty" merit functions that have been used by Eason [4] are:

$$M(x_i) = F(x_i) + 10^{20} \sum_{j=1}^{J} A_j \left| G_j(x_i) \right|,$$

where

$$A_j = \begin{cases} 1 \text{ if } G_j(x_i) < 0 \\ 0 \text{ if } G_j(x_i) \geq 0 \end{cases}$$

and

$$M(x_i) = F(x_i) + W \sum_{j=1}^{J} A_j [G_j(x_i)]^2.$$

Mischke [15] suggests the use of a function of the form:

$$M(X_i) = F(X_i) + \sum_{j=1}^{N} b \, [X_i(\text{good}) - x_i]^2$$

where

$$b = \begin{cases} 1 \text{ of the constraints are violated} \\ 0 \text{ of the constraints are not violated} \end{cases}$$

and $X_i(\text{good})$ represents a vector corresponding to a design that does not violate the constraints. The penalty function part of this merit surface is a smooth, unimodal, second-order hypersurface having its apex placed at a merit value of zero. The apex is located at the known point "$X_i(\text{good})$". This penalty surface provides gradients that direct all stray search excursions back into the feasible design space.

Example 7-2

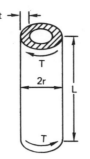

The proposed power shaft concept shown in the accompanying figure is to be used in one of the subassemblies of an aerospace system. The shaft is 10 cm long and must carry a maximum torque load of T = 50 Nm. Since the weight of every component of the total system is extremely important, it is desired to provide a design concept that will have adequate strength with minimum weight. This calls for the use of a material with an exceptional strength-to-weight ratio. The material selected for the design is the titanium alloy Ti-8Al-1Mo-1V, which has the following properties:

density $\qquad$ = 437. kg/m³

elastic modulus E $\quad$ = 1.21 x 10¹¹ N/m² $\quad$ (Pa)

shear stress yield
$\quad$ point S_{syp} $\qquad$ = 4.14 x 10⁸ N/m² $\quad$ (Pa)

For practical reasons dictated by the functional geometry, the radius "r" is limited to a maximum value of 2.0 cm. For practical reasons dictated by the manufacturing method, the tube thickness "t" must be no smaller than 1.0 mm.

If the factor of safety for the design is N = 1.5, specify the shaft radius and wall thickness for the best possible design.

For this power shaft, the modes of failure will be by shear stress and by shear buckling. The uniform shear stress generated by the applied torque load will be:

$$S_s = \frac{T}{2\pi r^2 t}.$$

In order to have a design that is safe from shear stress failure, the maximum shear stress theory of failure requires:

$$\frac{T}{2\pi r^2 t} \leq \frac{S_{syp}}{N}.$$

In order to have a design that is safe from shear buckling, the applied stress must not exceed the critical buckling stress:

$$\frac{T}{2\pi r^2 t} \leq \frac{S_{scr}}{N}.$$

Roark† reports that the critical buckling stress is:

$$S_{scr} = E\left[\frac{t}{L}\right]^2 \left\{ 3.0 + \sqrt{3.4 + 0.24\left[\frac{L}{\sqrt{tr}}\right]^3} \right\}$$

if

$$\frac{L}{r} \leq 7.72 \ (\frac{r}{t})^{\frac{1}{2}}$$

and

$$S_{scr} = 0.272 \ E \ (\frac{t}{r})^{3/2}$$

if

$$\frac{L}{r} > 7.72 \ \left[\frac{r}{t}\right]^{\frac{1}{2}} .$$

Thus the problem may be formulated in standard form as:

The design variables are: r and t.

The merit function to be minimized is the weight:

$W = 2\pi rtL\rho.$

The regional constraints are:

for shear stress: $\dfrac{T}{2\pi r^2 t} \leq \dfrac{S_{syp}}{N}$

for shear buckling: $\dfrac{T}{2\pi r^2 t} \leq \dfrac{S_{scr}}{N}$

for the given geometry:

$$.001 \leq t \leq r \leq .02m$$

The following program which solves this optimization problem uses subroutine SEEK1 from the OPTISEP [18] package. This routine utilizes the Hooke and Jeeves pattern search algorithm. The subroutine UREAL defines the merit function, and a penalty function is automatically supplied by SEEK1 routine if any of the constraints are violated. The subroutine CONST defines the five inequality constraints in a form that requires the five PHI(I) values to be positive when the constraints are satisfied. In order to provide a suitable starting value for the search process, the pair of values r = 0.019 and t = 0.017 were selected because they were found not to violate any of the constraints. The subroutine ANSWER prints the final value of design variables, the final merit value, and the constraint equations.

```
C     * * * * * * * * * * * * * * * * * * * * * * * * * *
C     *    THIS PROGRAM PERFORMS THE OPTIMUM DESIGN     *
C     *    OF AN AEROSPACE POWER SHAFT FOR MINIMUM      *
C     *    WEIGHT.                                      *
C     *                        T. E. SHOUP  9/14/77     &
C     * * * * * * * * * * * * * * * * * * * * * * * * * *
C
      DIMENSION X(2),PHI(5),PSI(1),RMAX(2),RMIN(2),XSTRT(2),
     &          WORK1(2),WORK2(2),WORK3(2),WORK4(2)
      REAL L
C
      RMAX(1) = 0.02
      RMIN(1) = 0.001
```

†Roark, R. J. Formulas for Stress and Strain. New York: McGraw-Hill Book Co., 1965, p. 359.

```
                RMAX(2) = 0.02
                RMIN(2) = 0.001
C
C       SET STARTING VALUES
                XSTRT(1) = 0.019
                XSTRT(2) = 0.017
C
C       PERFORM THE OPTIMIZATION BY HOOKE AND JEEVES DIRECT SEARCH
C
                CALL SEEK1(2, RMAX, RMIN, 5, 0, 0.01, 0.01, XSTRT, 2,
              &         100, 300, 0, 1, X, WEIGHT, PHI, PSI, WORK1,
              &         WORK2, WORK3, WORK4)
C
C       WRITE THE ANSWERS
C
                CALL ANSWER (WEIGHT, X, PHI, PSI, 2, 5, 0)
C
                STOP
                END



C       * * * * * * * * * * * * * * * * * * * * * * * * * * * * * * * *
C       *                S U B R O U T I N E   U R E A L              *
C       *        THIS SUBROUTINE PRODUCES THE POWER SHAFT WEIGHT      *
C       *        GIVEN THE DIMENSIONS.                                *
C       *                                      T.E. SHOUP 9/14/77     *
C       * * * * * * * * * * * * * * * * * * * * * * * * * * * * * * * *
C
                SUBROUTINE UREAL (X, WEIGHT)
                DIMENSION X(1)
                REAL L
C
                RHO = 437.
                L = 0.1
                R = X(1)
                T = X(2)
                PI = 3.1415926
                WEIGHT = 2. + PI * R * T * L * RHO
C
                RETURN
                END



C       * * * * * * * * * * * * * * * * * * * * * * * * * * * * * * * *
C       *                S U B R O U T I N E   C O N S T             *
C       *        THIS SUBROUTINE DEFINES THE INEQUALITY CONSTRAINTS   &
C       *        ASSOCIATED WITH THE DESIGN.                          *
C       *                                      T.E. SHOUP  9/14/77    *
C       * * * * * * * * * * * * * * * * * * * * * * * * * * * * * * * *
C
                SUBROUTINE CONST (X,NCONS, PHI)
                DIMENSION X(1), PHI(1)
                REAL L
C
                E = 1.21E11
                SSYP = 4.14E8
                FS = 1.5
                TOR = 50.
                L = 0.1
                R = X(1)
                T = X(2)
                PI = 3.1415926
```

```
C
C     CALCULATE THE SHEAR STRESS CONSTRAINT
      PHI(1) = SSYP / FS - TOR / ( 2. * PI * R * R * T )
C
C     CALCULATE THE SHEAR BUCKLING CONSTRAINT
      SSCR = E * T * T / ( L * L ) * ( 3. + SQRT ( 3.4 + 0.24 * ( L /
     &SQRT ( T * R ) ) ** 3 ) )
      IF (L/R .GT. 7.72*SQRT(R/T)) SSCR = 0.272*E*(T/R)**1.5
      PHI(2) = SSCR / FS - TOR / ( 2. * PI * R * R * T )
C
C     CALCULATE THE GEOMETRIC CONSTRAINTS
      PHI(3) = 0.02 - R
      PHI(4) = T - 0.001
      PHI(5) = R - T
C
      RETURN
      END
```

A portion of the output of this program now follows. For the solution found, it is clear that none of the constraints have been violated although a few of the constraints are very close to having a zero value. This indicates that the optimum value is likely on the boundary defined by these constraints.

```
            OPTIMUM SOLUTION FOUND
          MINIMUM  U =  0.20010402E 01

                X( 1) =  0.38074213E-02
                X( 2) =  0.19900010E-02
INEQUALITY CONSTRAINTS
                PHI( 1) =  0.14873139E 06
                PHI( 2) =  0.80149512E 10
                PHI( 3) =  0.16192579E-01
                PHI( 4) =  0.99000101E-03
                PHI( 5) =  0.18174202E-02
```

7.12 Indirect Optimization

No treatment of multidimensional optimization techniques would be complete without a discussion of the calculus of stationary points.

In order for a multidimensional function to have a maximum, a minimum, or a saddle point, it is necessary that all first derivatives with respect to each of the "N" independent variables be zero. Thus, for the function "Merit = $F(x_1, x_2, \ldots, x_N)$," a stationary point will satisfy:

$$\frac{\partial F}{\partial x_1} = 0, \quad \frac{\partial F}{\partial x_2} = 0, \quad \ldots, \quad \frac{\partial F}{\partial x_N} = 0$$

In order to determine whether a stationary point is a minimum, a maximum, or a saddle point, it is necessary to examine the second derivatives of the function. A convenient way to describe the nature of the second derivatives is by means of the Hessian matrix, which is of the form:

195

$$\text{Hessian} = \begin{bmatrix} \dfrac{\partial^2 F}{\partial x_1^2} \,, & \dfrac{\partial^2 F}{\partial x_1 \partial x_2} \,, & \cdots & \dfrac{\partial^2 F}{\partial x_1 \partial x_N} \\[4mm] \dfrac{\partial^2 F}{\partial x_2 \partial x_1} & \cdots & & \\[2mm] \cdot & & & \\ \cdot & & & \\ \cdot & & & \\[2mm] \dfrac{\partial^2 F}{\partial x_N \partial x_1} & \cdots & & \dfrac{\partial^2 F}{\partial x_N^2} \end{bmatrix}$$

A necessary and sufficient condition for a stationary point to be a local minimum is that its Hessian matrix be positive definite. This means that all of its eigenvalues will be positive. A necessary and sufficient condition for a stationary point to be a local maximum is that its Hessian matrix be negative definite. This means that all of its eigenvalues will be negative.

One way to mechanize this information is shown in Figure 7-13.

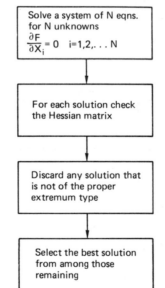

Figure 7-13   The indirect method of optimization.

First, the system of equations corresponding to the "N" first partial derivatives is found. This system must be solved for all possible sets of design variables that satisfy the "N" equations. If these equations are linear, the problem is straightforward because only one solution set will exist. If the system is nonlinear, as is most often the case, there may be many solution sets. Once the solutions have been isolated, the designer must discard all of the solution sets that are not of the desired extremum type. This requires a check of the second partial derivatives evaluated at each of the solution design points. Once the solution sets have been reduced to a final group, the designer must check to see which of the group has the most desirable merit value. This one will be declared the optimum.

Although the foregoing technique does seem mathematically straight-forward, it is, in reality, not extremely practical, because the optimum in many design situations will occur at a boundary rather than at a stationary point. The technique does point out the need for methods to extract eigenvalues and for methods to solve systems of nonlinear algebraic equations.

One interesting extension of the technique of stationary points is the method of Lagrange multipliers. This technique allows equality constraints of the form

$$Q_1 (x_1, x_2, \ldots, x_N) = 0$$
$$\cdot$$
$$\cdot$$
$$\cdot$$
$$Q_J (x_1, x_2, \ldots, x_N) = 0$$

to be satisfied in the optimization process. To facilitate the solution of this problem, a new merit function must be formed that is a linear combi-nation of the old merit function and each of the constraint equations multiplied by a unique constant. This new merit function will be:

$$M(x_i, \lambda_j) = F(x_i) + \lambda_1 Q_1 + \lambda_2 Q_2 + \ldots \lambda_J Q_J.$$

The λ_j values are called Lagrangian multipliers and are to be treated as additional unknowns to be determined in the solution process. Thus the system used to locate stationary points consists of $J + N$ equations and $J + N$ unknowns:

$$x_1, x_2, \ldots, x_N, \lambda_1, \lambda_2, \ldots, \lambda_J.$$

If each of the constraints is satisfied, the additional λ terms each contri-bute nothing to the new merit function. In this case the optimization of "M" is equivalent to the optimization of "F". It should be noted that, in the equations to be solved for the stationary point, the partial derivatives of the new merit function with respect to the unknown Lagrangian multi-pliers revert to the constraint equations. Thus:

$$\frac{\partial M}{\partial \lambda_1} = Q_1 = 0$$

$$\frac{\partial M}{\partial \lambda_2} = Q_2 = 0 \qquad \text{etc.}$$

7.13 Considerations in the Selection of a Multidimensional Optimization Algorithm

Every complex design problem has its own unique characteristics that make it a special challenge to the traditional optimization algorithms. Although no single method can hope to be universally successful, some techniques are better suited for certain types of problems. Careful selection of an appropriate algorithm class can often cut down on the waste of computer time and designer effort. In choosing an appropriate algorithm, certain practical matters should be considered. These are listed in this section in an order that implies neither priorities nor order of importance.

1. Consider the Nature of the Merit Surface. If it is known, the shape characteristics of a merit surface can often give considerable insight into the selection or elimination of an algorithm class. For example, even if only mild ridges are present, the sectioning and gradient methods should not be used. If strong ridges are present, the deflected gradient methods should also be eliminated in favor of pattern search methods. If the merit surface has sharp valleys, the simplicial method or Rosenbrock's method will often outperform the method of Hooke and Jeeves. If multimodality is suspected, it is wise to choose several diverse starting points in the design space to see if every start leads to the same solution. If more than one local optimum is found, the best may be selected for use as a design. Unfortunately, even the most thorough and diverse selection of starting points will not guarantee the designer that he or she will discover every local optimum.

2. Consider the Nature of the Design Variables. Although most engineering quantities come in infinitely variable sizes, a few engineering problems do deal with discrete-valued or integer-valued variables. Examples of these quantities include such things as the size of a pipe, the number of teeth on a gear, or the number of bolts on a flange. Whenever discrete-valued problems occur, they may be treated by traditional algorithms just as if they were composed entirely of variable quantities. When an optimum solution is achieved, the designer will need to round-off each variable size to a suitable neighboring value. This process often involves a check of the merit associated with both rounding up and rounding down. If the problem is sufficiently complex, it may be that this approach will not give the best design. In such cases it may be necessary to employ techniques that are designed to handle this class of problems. Such techniques are available in the optimization literature. Unfortunately they carry no guarantee of better performance than the variable "round-off" method.

3. Consider the Time Required to Solve a Problem. The time required to utilize a given algorithm to solve an optimization problem is composed of the preparation time and the computer run time. An extra coding (such as a penalty function) will add to the cost of using an algorithm. Often the engineering designer is faced with a trade-off in his or her decision to use a particular method. Should extra programming effort to prepare the problem for a "fast"

198

method be spent or should the designer opt for the slow but easy method. Only the designer can make this decision, because the cost of programming time and the cost of computer time will vary from place to place.

4. Consider the Nature of the Algorithm. The nature of an algorithm will often give a clue as to its suitability for a particular problem. Methods that require analytical derivatives should be avoided if possible, because the derivatives of a merit function are often impossible to obtain (for example, if the merit function depends on experimental data). In addition, if analytical derivatives are of high complexity, differentiation can greatly add to the preparation cost and can also increase the opportunity for human error. Methods that feature built-in problem scaling generally have increased efficiency and utility. It is interesting to note that computer codes having great versatility need not be large in size or slow in operation. Because it is always desirable to try several algorithms in a given problem, a "package" approach is highly desirable. The use of a "package" strategy allows the designer to interchange algorithms simply by changing one subroutine call in the main program. Thus considerable time and effort is saved by eliminating the need for reprogramming.

7.14 FORTRAN Codes for Multidimensional Optimization

A large number of excellent FORTRAN subroutines for multidimensional optimization are presently in existence, and new ones are constantly being generated. To attempt to mention all of these would inevitably lead to the omission of someone's favorite. To avoid this pitfall, a group of 15 have been listed in Table 7-2. These 15 algorithms were selected because:

1. They all are commonly available.

2. They all are well documented.

3. They all have been in common use for some time.

4. They represent a diversity of methods including at least one "package" set of routines.

A similar group of FORTRAN codes containing many of those listed here was the topic of a recent comparative study by Eason [4, 5]. In this investigation a group of FORTRAN codes was applied to 10 diverse problems of moderate complexity. The relative performance of each code was evaluated in terms of generality, efficiency, and cost to prepare and run. Although the results of Eason should be interpreted with care, he does have a number of favorable comments about the success of the single routine PATSH and about the desirability of the "package" of routines OPTISEP [18].

Table 7-2 FORTRAN Codes for Multidimensional Optimization

Name	Source Reference	Method	Special Features
ADRANS	18	Local random search followed by a pattern search	-
CLIMB	19	Rosenbrock	Contains a built-in penalty function.
DAVID	18	Davidon-Fletcher-Powell	Contains secant approximation to derivatives.
DFMCG	13	Fletcher-Reeves	User must supply derivatives & penalty function.
DFMFP	13	Davidon-Fletcher-Powell	User must supply derivatives & penalty function.
FMIND	14	Hooke & Jeeves Pattern Search	User must supply penalty function.
GRAD4	15	Steepest descent method	User must supply penalty function.
GRID4	15	Grid and star network search	User must supply penalty function.
MEMGRD	18	Extension of Davidon-Fletcher-Powell	Step size is based on retained information from previous step.
NMSERS	19	Nelder & Mead	User must supply penalty function.
PATSH	20	Modified Hooke & Jeeves	Stepsize proportional to magnitude of variables and stepsize growth multiplier.
RANDOM	18	Random search with shrinkage	Built-in penalty function.
SEEK1	18	Hooke & Jeeves pattern search followed by a random search	-
SEEK3	18	Sequential Hooke & Jeeves	-
SIMPLX	18	Sequential simplex method	-

Problems

7.1 A waste-treatment settling tank as shown in the accompanying figure is to be designed to hold 40,000 liters of liquid waste. The tank is made of 10 cm concrete with reinforcing. Design the tank so that a minimum amount of concrete is used.

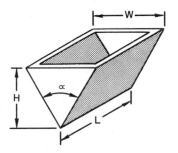

7.2 Solve Problem 7.1 if the container has a closed top.

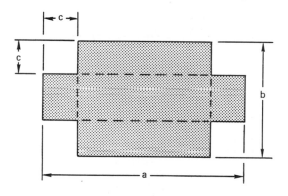

7.3 A container manufacturer is designing an open top container of sheet material as shown in the accompanying figure. The sheet is to be cut, folded on the dashed lines, and then welded along the four seams. If the area of the box cannot exceed 1.0 m² and if no dimension (a,b,c) can be larger than three times any other dimension, what size container will have a maximum volume?

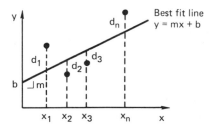

7.4 It is often convenient to be able to find the best-fit straight line for an arbitrary set of experimental data points as indicated in the accompanying figure. The best fit straight line is expressed as: $y = mx + b$. The accumulated error for "k" data points is expressed as:

$$\text{SUM} = \sum_{i=1}^{k} [d_i^2]^{j/2}$$

Write a computer program that will perform the minimization of SUM to find the optimum values of m and b for:

(a) $j = 1$ (absolute value of error)

(b) $j = 2$ (least squares error)

(c) $j = 4$ (fourth power of error)

for the data:

i	x_i	y_i
1	1.0	1.0
2	1.5	2.8
3	2.8	2.5
4	2.7	4.0
5	4.3	3.0

7.5 If Problem 7.4 is solved using the best fit of a second order equation: $y = a_1 x^2 + a_2 x + a_3$, find the optimum values for a_1, a_2 and a_3 for the three cases.

7.6 Use two different optimization techniques to solve Example 6-1 from the previous chapter.

7.7 It is desired to choose the dimensions for the slider crank mechanism show in the accompanying figure. It is desired to do this to provide the best approximation (minimum error) for the function:

$$X_{desired} = 1 + \cos^2\theta \qquad \text{for } 0° \leq \theta \leq 90°.$$

The actual input-vs-output behavior for the device is:

$$X_{actual} = \cos(\theta + \theta_0) + \sqrt{\cos^2(\theta + \theta_0) - (A_1^2 - A_2^2)}$$

You may express the error as:

$$\text{Error} = \int_0^{90°} (X_{desired} - X_{actual})\, d\theta.$$

The particular design requires that:

$$A_1 < A_2$$

$$0.1 \leq A_1 \leq 10$$

$$0.1 \leq A_2 \leq 10$$

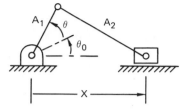

7.8 For the gear train shown in the accompanying figure, the speed ratio that relates the angular velocity of the last gear to that of the first gear depends on the number of teeth (N_i) on each gear in the train:

$$\text{speed ratio} = \frac{N_1 N_3 N_5}{N_2 N_4 N_6}.$$

It is desired to design a gear train that will have a speed ratio that is as close to $1/(10\pi)$ as possible. For practical reasons the number of teeth on each gear must be restricted to:

$$20 \leq N_i \leq 100 \qquad i = 1,2,..,6.$$

Find the best possible design by specifying the values of N_1, N_2, N_3, N_4, N_5, and N_6.

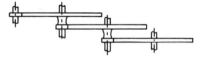

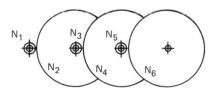

7.9 Although it is rather simple mathematically, the Rosenbrock merit function:

$$y = 100(x_2 - x_1^2)^2 + (1 - x_1)^2$$

contains a curved valley as shown in the accompanying figure. The minimum merit location lies at the point (1.0, 1.0); however, if a starting value in the second quadrant is selected, convergence can sometimes be difficult to achieve. Start with the point (-1.,1.) and try this optimization problem by:

(a) the sectioning method,
(b) a gradient method, and
(c) a pattern search method.

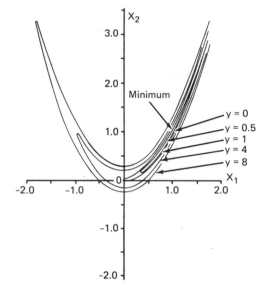

7.10 A 8.9-cm OD, 7.8-cm ID steel (k = 43.27 W/m°C) pipe, shown in the accompanying figure, transports a fluid at 148°C. The pipe is insulated with two layers of insulation as shown. The inner layer of insulation has a thermal conductivity of k = .20 W/m°C and the outer layer has a thermal conductivity of k = 0.5 W/m°C. The coefficient of convection on the inside of the pipe is 230 W/m² °C and on the outside of the outer insulation it is 23 W/m² °C. Design the insulation layers for minimum cost if the maximum outer surface temperature allowed is 38°C and the ambient air temperature is 27°C. The maximum allowable outer radius is 12 cm. The cost of the inner insulation material is $35/m³, and the cost of the outer material is $100/m³. (You may neglect the thermal contact loss between layers of material.)

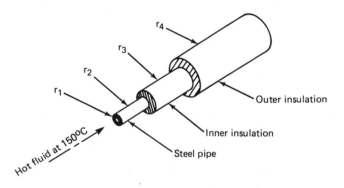

References

1. Brooks, S. H. "A Discussion of Random Methods for Seeking Maxima." Operations Research 6 (Mar. 1958): 244-51.

2. Carroll, C.W. "An Operations Research Approach to the Economic Optimization of a Kraft Pulping Process." Doctoral dissertation, The Institute of Paper Chemistry, Appleton, Wisc., 1959.

3. Carroll, C. W. "The Created Response Surface Technique for Optimizing Nonlinear Restrained Systems." Operations Research, 9 (1961): 169-184.

4. Eason, E. D., and Fenton, R. G. "A Comparison of Numerical Optimization Methods for Engineering Design." ASME Paper 73-DET-17, 1974.

5. Eason, E. D., and Fenton, R. G. "Testing and Evaluation of Numerical Methods for Design Optimization." UTME-TP 7204, University of Toronto, Sept. 1972.

6. Fiacco, A. V., and McCormick, G. P. "Computational Algorithm for the Sequential Unconstrained Minimization Technique for Nonlinear Programming." Management Science 10 (1964): 601-17.

7. Fiacco, A. V., and McCormick, G. P. Nonlinear Sequential Unconstrained Minimization Techniques. New York: John Wiley and Sons, 1968.

8. Fiacco, A. V., and McCormick, G. P. "The Sequential Unconstrained Minimization Technique for Nonlinear Programming, A Primal Dual Method." Management Science 10 (1964): 360-66.

9. Fletcher, R., and Powell, M. J. D. "A Rapidly Convergent Descent Method for Minimization." Computer J. 6 (1963): 163-68.

10. Fletcher, R., and Reeves, C. M. "Function Minimization by Conjugate Gradients." Computer J. 7 (1964): 149-154.

11. Flectcher, R., ed. Optimization. New York: Academic Press, 1969.

12. Hooke, R., and Jeeves, T. A. "Direct Search Solution of Numerical and Statistical Problems." J. Assoc. Comp. Mach. 8 (1961): 212-229.

13. IBM System/360 Scientific Subroutine Package (SSP). 360A-CM-03X Version 3, 6th ed. Mar. 1970.

14. MIT Information Processing Center. Applications Program Series AP-78 (APD-46), Cambridge, Mass. 02139.

15. Mischke, C. R. An Introduction to Computer-Aided Design. Englewood Cliffs, N.J.: Prentice-Hall, Inc., 1968. (Mechanical Engineering Department, Iowa State University, Ames, Iowa 50010.).

16. Nelder, J. A., and Mead, R. "A Simplex Method for Function Minimization." Computer J. 7 (1964): 308-313.

17. Rosenbrock, H. H. "An Automatic Method for Finding the Greatest or Least Value of a Function." Computer J. 3 (1960): 175-184.

18. Siddall, J. N. "OPTISEP" Designers' Optimization Subroutines (ME/71/DSN/REP1). Faculty of Engineering, McMaster University, Hamilton, Ontario, Canada, 1971.

19. WATLIB. University of Toronto Computer Center, Toronto 5, Ontario, Canada.

20. Whitney, D. E. "Joint Mechanical and Civil Engineering Computing Facility." MIT, 77 Massachusetts Ave., Cambridge, Mass. 02139.

21. Wilde, D. J., and Beightler, C. S. _Foundations_ _of_ _Optimization_. Englewood Cliffs, N.J.: Prentice-Hall, Inc., 1967.

8 MISCELLANEOUS COMPUTER METHODS

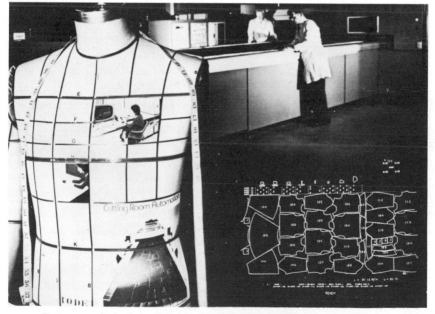

For ease of design changes, the computer forms an integral part of the pattern layout and cutting equipment in this application from the apparel industry. [Photo courtesy of Hughes Aircraft Co.]

8.1   Introduction

Traditionally, engineering design activities provide considerable quantitative output.   It is this quantitative nature of engineering problems that leads to the processing challenges of numerical information.   Indeed, the key to understanding for many problems is often revealed as a result of a thoughtful presentation of the numerical data describing the problem. Conversely, poor data presentation can often lead to confusion and misinterpretation and can increase the opportunity for human error.   In the manipulation and management of engineering data, certain basic computational tools have become extremely useful.   Among these are:

1.   numerical interpolation;
2.   curve fitting;
3.   numerical differentiation; and
4.   numerical integration.

It is the purpose of this chapter to discuss these topics within the context of engineering data management.

8.2   Interpolation

Engineering data is frequently available in tabular form.   This method of data representation arises due to the fact that the data was obtained only for discrete values by experimental means or due to the fact that practical limitations on the volume of data to be managed requires that only a few values be stored.   The process of interpolation is that of finding a value of a function at some intermediate location in the tabular data.

The simplest form of interpolation is linear interpolation based on a linear approximation through points (x_k, y_k) and (x_{k+1}, y_{k+1}) as shown in Figure 8-1.   The equation of the line is:

$$\frac{y - y_k}{x - x_k} = \frac{y_{k+1} - y_k}{x_{k+1} - x_k}$$

or

$$y = \frac{y_k(x - x_{k+1}) - y_{k+1}(x - x_k)}{x_k - x_{k+1}}.$$

Thus, if one knows two tabular values that surround a given x value, this linear relationship can be used to find an approximation to the corresponding y value.   It is generally accepted that the use of more neighboring points and an approximation of more complexity than a line will give

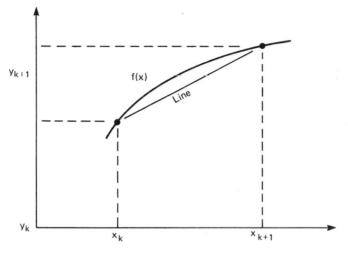

Figure 8-1 Linear interpolation.

better results. In this section methods are developed to find a unique n^{th}-order polynomial $P_n(x)$, which approximates the function $f(x)$ by satisfying all "n + 1" tabular points (x_i, y_i), $i = 0, 1, \ldots, n$. The polynomial is said to have constraints:

$$P_n(x_i) = y_i \qquad i = 0, \ldots, n.$$

The methods of finding this polynomial fall into three categories: Lagrangian methods, difference methods, and iterative methods.

Lagrange Interpolation

In Lagrangian interpolation, n + 1 table values (x_i, y_i), $i = 0, \ldots, n$, are given representing points on the function $y = f(x)$ on the interval $x_0 \leq x \leq x_n$. For this method the interpolating polynomial will be expressed as:

$$P_n(x) = Y_0 b_0(x) + y_1 b_1(x) + \ldots + y_n b_n(x),$$

where each $b_j(x)$ is a polynomial of degree "n". These polynomials can be determined by imposing the n + 1 constraint equations:

$$P_n(x_i) = y_i \qquad i = 0, \ldots, n.$$

This will give a system of the form:

209

$$y_0 b_0(x_0) + y_1 b_1(x_0) + \ldots + y_n b_n(x_0) = y_0$$
$$\vdots$$
$$y_0 b_0(x_n) + \ldots \ldots + y_n b_n(x_n) = y_n$$

If the $b_j(x_i)$ values are chosen so that:

$$b_j(x_i) = \begin{cases} 1 & i = j \\ 0 & i \neq j \end{cases}$$

then the constraint equations will be satisfied. This condition requires that each $b_j(x)$ has zeros corresponding to every value of x except x_j. Thus a general $b_j(x)$ polynomial will be:

$$b_j(x) = C_j(x - x_0)(x - x_1)\ldots(x - x_{j-1})(x - x_{j+1})\ldots(x - x_n).$$

Because

$$b_j(x_j) = 1,$$

the coefficient C_j can be found to be

$$C_j = \frac{1}{(x_j - x_0)(x_j - x_1)\ldots(x_j - x_{j-1})(x_j - x_{j+1})\ldots(x_j - x_n)}.$$

Thus the final polynomial will be:

$$P_n(x) = \sum_{j=0}^{n} y_j \frac{(x - x_0)(x - x_1)\ldots(x - x_{j-1})(x - x_{j+1})\ldots(x - x_n)}{(x_j - x_0)(x_j - x_1)\ldots(x_j - x_{j-1})(x_j - x_{j+1})\ldots(x_j - x_n)}$$

This polynomial can be written in an easier form using:

$$L_j(x) = (x - x_0)(x - x_1)\ldots(x - x_{j-1})(x - x_{j+1})\ldots(x - x_n).$$

Thus

$$P_n(x) = \sum_{j=0}^{n} y_j \frac{L_j(x)}{L_j(x_j)}$$

The Method of Divided Differences

Although there are a number of difference-based formulations for interpolation, the Newton's forward-difference formulation (also known as the Newton Gregory formulation) is the most commonly used. It is based on an interpolating polynomial of the form:

$$P_n(x) = c_0 + c_1(x - x_0) + c_2(x - x_0)(x - x_1) + \ldots +$$
$$c_n(x - x_0)(x - x_1)\ldots(x - x_{n-1}).$$

210

The coefficients c_j can be found by applying the constraint equations:

$$P_n(x) = y_i \qquad i = 0, \ldots, n.$$

This would give a system of equations of the form:

$$c_0 = y_0$$

$$c_0 + c_1 (x_1 - x_0) = y_1$$

$$c_0 + c_1 (x_2 - x_0) + c_2 (x_2 - x_0)(x_2 - x_1) = y_2$$

$$\vdots$$

$$c_0 \ldots + c_n(x_n - x_0)(x_n - x_1)\ldots(x_n - x_{n-1}) = y_n$$

This form fortunately is linear and triangular. The values of c_j can be found without difficulty; however, there is an easier way to find them based on forward finite differences. If the x values are equally spaced:

$$x_{i+1} - x_i = h,$$

then in general $x_i = x_0 + ih$ for $i = 1, \ldots, n$. When this expression is used, the equations to be solved are:

$$y_0 = c_0$$

$$y_1 = c_0 + c_1 h$$

$$y_2 = c_0 + c_1(2h) + 2h^2 c_2$$

$$\vdots$$

$$y_i = c_0 + c_1 ih + c_2 ih[(i-1)h] + \ldots + c_i (i!)h^i$$

If one solves for the coefficients, the result is:

$$c_0 = y_0$$

$$c_1 = \frac{y_1 - c_0}{h} = \frac{y_1 - y_0}{h} = \frac{\Delta y_0}{h} .$$

In this expression, Δy_0 is called the first forward difference. As the process continues:

$$c_2 = \frac{1}{2h^2} [y_2 - c_0 - 2hc_1] = \frac{1}{2h^2} [(y_2 - y_1) - (y_1 - y_0)] = \frac{1}{2h^2} [\Delta(\Delta y_0)]$$

$$= \frac{\Delta^2 y_0}{2h^2} .$$

In this expression, $\Delta^2 y_0$ is called the second forward difference because it is the difference of the differences. In general, the c_j coefficients of the polynomial can be expressed as:

$$c_j = \frac{\Delta^j y_0}{(j!)h^j} .$$

In general, the higher order differences of the function $y = f(x)$ are defined over the interval $x_0 \le x \le x_n$ as:

$$\Delta^j y_i = \Delta^{j-1} y_{i+1} - \Delta^{j-1} y_i \qquad i = 0, \ldots n - j.$$

These differences frequently are tabulated as shown in Table 8-1.

Table 8-1 Forward Differences

x_i	y_i	Δy_i $=$ $y_{i+1} - y_i$	$\Delta^2 y_i$ $=$ $\Delta y_{i+1} - \Delta y_i$	$\Delta^3 y_i$ $=$ $\Delta^2 y_{i+1} - \Delta^2 y_i$	$\Delta^4 y_i$ $=$ $\Delta^3 y_{i+1} - \Delta^3 y_i$	$\Delta^5 y_i$ $=$ $\Delta^4 y_{i+1} - \Delta^4 y_i$
x_0	y_0					
		Δy_0				
x_1	y_1		$\Delta^2 y_0$			
		Δy_1		$\Delta^3 y_0$		
x_2	y_2		$\Delta^2 y_1$		$\Delta^4 y_0$	
		Δy^2		$\Delta^3 y_1$		$\Delta^5 y_0$
x_3	y_3		$\Delta^2 y_2$		$\Delta^4 y_1$	$\vdots$
		Δy_3		$\Delta^3 y_2$	$\vdots$	
x_4	y_4		$\Delta^2 y_3$	$\vdots$		
		Δy_4	$\vdots$			
x_5	y_5	$\vdots$				
$\vdots$	$\vdots$					

In this table the differences of any given order depend on the differences of the next lower order. To illustrate the use of the Newton forward-difference interpolation method, an example application is considered.

Example 8-1

Suppose that a table of information is given:

x_i	y_i
10.	0.17365
20.	0.34202
30.	0.50000
40.	0.64279
50.	0.76604
60.	0.86603

[This data corresponds to the function $y = \text{six}(x_{degrees})$.]

Find the value of y at $x = 23$ using the method of divided differences.

Using this data a difference table can be constructed as

212

x_i	y_i	Δy_i	$\Delta^2 y_i$	$\Delta^3 y_i$	$\Delta^4 y_i$	$\Delta^5 y_i$
10.	0.17365	-				
		0.16837	-			
20.	0.34202		-0.01039	-		
		0.15798		-0.00480	-	
30.	0.50000		-0.01519		+0.00045	-
		0.14279		-0.00435		+0.00018
40.	0.64279		-0.01954		+0.0063	
		0.12325		-0.00372		
50.	0.76604		-0.02326			
		0.09999				
60.	0.86603					

The value of x_0 may be chosen anywhere in the table. Suppose that the value x_0 = 20. is chosen. The necessary differences lie on a diagonal down from x_0. As many or as few higher-order differences may be used as desired. In general, the accuracy will improve if more differences are used. One advantage to this method is that it allows the user to add differences to a previous calculation without starting over in the calculation process. Thus, if one does not know how many terms to carry along, one can add terms until the contribution of the added terms is so small that the number of decimal digits has stabilized. In the case of this example problem, h = 10. Using only first-order differences gives:

$$y(23) = y + \frac{\Delta y_0}{h} (23. - x_0) = .34202 + \frac{.15798}{10} \quad (3)$$

$$= 0.38941.$$

Using the first- and second-order differences gives:

$$y(23.) = 0.38941 + \frac{\Delta^2 y_0}{h} (23 - x)(23 - x_1) = 0.38941 +$$

$$\frac{(-0.01519(3)(-7)}{200.} = 0.39100$$

Using first-, second- and third-order differences gives:

$$y(23.) = 0.39100 + \frac{\Delta^3 y_0}{6h^3}(23 - x)(23 - x_1)(23 - x_2)$$

$$= 0.39074.$$

Clearly, this answer is approaching the actual value 0.39073.

Other difference forms can be applied to achieve alternative interpolation schemes. Among these methods are the Newton's backward-difference formulation, Gauss's forward-difference formulation, and Gauss's backward-difference formulation.

Iterative Interpolation Methods
Iterative interpolation schemes are based on the repeated application of a simple interpolation process. The best-known of these methods is

Aitken's method, which is based on repeated linear interpolation. This method is now described.

It was shown previously that a linear interpolation between points (x_0, y_0) and (x_i, y_i) will be:

$$y_{i1}(x) = \frac{1}{x_i - x_0} [y_0(x_i - x) - y_i(x_0 - x)].$$

Using this relationship, a table of values $y_{i1}(x)$ i = 1, ..., n can be generated for a given x value. Using these values and linear interpolation of the form:

$$y_{i2}(x) = \frac{1}{x_i - x_1} [y_{11}(x)(x_i - x) - y_{i1}(x)(x_1 - x)],$$

a new family of relationships can be found. It can be shown by simple substitution that the $y_{i2}(x)$ relationships are second-order polynomials that satisfy the three points (x_0, y_0), (x_1, y_1), and (x_i, y_i). Once the family of polynomials y_{i2} has been found, linear interpolation using the values of $y_{i2}(x)$ may be accomplished to assemble:

$$y_{i3}(x) = \frac{1}{x_i - x_2} [y_{22}(x)(x_i - x) - y_{i2}(x)(x_2 - x)].$$

which is a third order polynomial that goes through the points (x_0, y_0), (x_1, y_1), (x_2, y_2) and (x_i, y_i). As the process continues, the values of $y_{ij}(x)$ tend to the value of $f(x)$. Although it is possible to extend this method beyond a third order polynomial, it is usually unwise to do so because of error propagation. It is important to note, however, that Aitken's process does not require uniformly spaced data values. To illustrate this method, Example 8-1 is now interpolated in Example 8-2 by Aitken's method to find y(23.).

Example 8-2

Suppose that it is desired to apply Aitken's method to solve the problem posed in Example 8-1. The following table of results is based on the repeated linear interpolation using x = 23. As the method progresses, the table values tend to approach the true value of 0.39073.

Numerical Results

i	x_i	y_i	y_{i1}	y_{i2}	y_{i3}
0	10	0.17365	-		
1	20	0.34202	0.39253	-	-
2	30	0.50000	0.38578	0.39051	-
3	40	0.64279	0.37694	0.39019	0.39073
4	50	0.76604	0.36618	0.38990	0.39072
5	60	0.86603	0.35367	0.38962	0.39072

Inverse Interpolation

Inverse interpolation is the process whereby one finds the value of x corresponding to a given function value y. In this case y is a value between two given values in the table of data. In such a situation, one could invert the table by interchanging the roles of x and y. The only disadvantage to this procedure is that the arguments of the table are no longer uniformly spaced. For this reason methods that are based on uniform spacing cannot be used.

8.3   Curve Fitting

In fitting tabular data with an approximating function, there are two basic schemes. The first scheme requires that the approximating function (perhaps a piecewise function) pass through every point in the table. The interpolation methods discussed in the previous section satisfy this requirement. The alternate approach is to find a simple function that applies over the total range of the table but does not exactly satisfy every data point. This category of problems is called curve fitting and seeks to minimize the error between the simple function and the tabular data values. The usual approach to this problem is to seek an approximating function such that the sum of the squares of the difference between the function and the actual data is a minimum. This procedure is known as the method of least squares.

Method of Least Squares

If one is given n + 1 data points $[(x_0, y_0) \ldots (x_n, y_n)]$ and it is desired to use an approximation function g(x) on the range

$$x_0 \leqq x \leqq x_n,$$

the functional error at any tabular point will be:

$$error_i = g(x_i) - y_i.$$

The sum of squares of these individual errors will be:

$$E = \sum_{i=0}^{n} [g(x_i) - y_i]^2.$$

It is customary to choose g(x) so that it is a linear combination of suitable terms of the form:

$$g(x) = c_1 g_1(x) + c_2 g_2(x) + \ldots + c_k g_k(x).$$

In order to have a minimum value for E:

$$\frac{\partial E}{\partial c_1} = \frac{\partial E}{\partial c_2} = \ldots \frac{\partial E}{\partial c_k} = 0$$

Because:

$$E = \sum_{i=0}^{n} [c_1 g_1(x_i) + c_2 g_2(x_i) + \ldots + c_k g_k(x_i) - y_i]^2,$$

the requirement of minimum error will give rise to the equation system:

$$\frac{\partial E}{\partial c_1} = 2\Sigma[c_1 g_1(x_i) + \ldots + c_k g_k(x_i) - y_i]g_1(x_i) = 0$$
$$\vdots$$
$$\frac{\partial E}{\partial c_k} = 2\Sigma[c_1 g_1(x_i) \ldots + c_k g_k(x_i) - y_i]g_k(x_i) = 0$$

Clearly, these k equations can be put into the form:

$$\begin{bmatrix} \Sigma g_1{}^2(x_i) & \Sigma g_1(x_i)g_2(x_i) & \ldots & \Sigma g_1(x_i)g_k(x_i) \\ \vdots & & & \\ \Sigma g_1(x_i)g_k(x_i) & \ldots & & \Sigma g_k{}^2(x_i) \end{bmatrix} \begin{bmatrix} c_1 \\ \vdots \\ c_k \end{bmatrix} = \begin{bmatrix} \Sigma g_1(x_i)y_i \\ \vdots \\ \Sigma g_k(x_i)y_i \end{bmatrix}$$

Since the coefficients in the matrix on the left and in the vector on the right can be determined from the tabular data, this system of k linear equations and k unknowns can be solved. As long as g(x) is linear in its coefficients, any reasonable function may be used. The actual selection should be based on individual judgment of the data. Special characteristics of the data that influence this judgment are periodicity, exponential or log tendencies, symmetry, and asymptotic behavior.

Although the procedure should be used with care, it is possible to break the tabular data into a few separate regions and then do a curve fit for each range. Such a procedure is often justified if the physical situation suggests that a transition from one domain to another has occurred. Examples of this situation in engineering problems include the transition between laminar and turbulent flow, the transition between subsonic and supersonic flow, and the transition between prebuckling and postbuckling behavior. The engineer who utilizes curve fitting should be careful not to use the approximation formula outside of the range of approximation.

Orthogonal Polynomials
If in formulating the approximation function the $g_i(x)$ functions are orthogonal polynomials such that:

$$\Sigma g_j(x_i)g_k(x_i) = 0 \qquad j \neq k,$$

216

the system of linear equations found previously will be simplified to a pure diagonal form. Thus the coefficients will be:

$$c_j = \frac{\sum\limits_{i=0}^{n} g_j(x_i)y_i}{\sum\limits_{i=0}^{n} g_j^2(x_i)}.$$

In view of the power of this simplification, it is not surprising that many computer curve-fitting subroutines rely on the use of orthogonal polynomials.

Spline Functions

Although spline functions are a recent mathematical tool, the physical basis for their development is well established in engineering drafting. A spline is a flexible strip or ruler that is constrained to pass through a given set of points $(x_i,\ y_i)$. When it is constrained in this way, the spline assumes a shape that minimizes its stored strain energy. Based on the assumption of small deflections, the theory of beam deflections can be used to demonstrate mathematically that the spline is a connected group of cubic polynomials arranged so that adjacent curves join each other with continuous first and second derivatives. Such functions are called cubic spline functions. In order to construct a cubic spline, it is necessary to specify the coefficients that uniquely describe each cubic polynomial between given data points. For example, in Figure 8-2 it is necessary to define each of the cubic functions $q_1(x)$, $q_2(x)$, $\ldots$ $q_m(x)$. In their most general form these polynomials will be:

$$q_i(x) = k_{1i} + k_{2i}x + k_{3i}x^2 + k_{4i}x^3 \qquad i = 1, 2, \ \ldots, \ m,$$

where the k_{ij} terms are constants to be determined by applying the constraint conditions mentioned previously.

The first 2m constraint conditions are that the splines must join each other at the data points. Thus:

$$q_i(x_i) = y_i \qquad\qquad i = 1, \ \ldots, \ m$$

$$q_{i+1}(x_i) = y_i \qquad\qquad i = 0, \ \ldots, \ m - 1.$$

The next 2m - 2 conditions are that the first and second derivatives must match at all internal data points. Thus:

$$q'_{i+1}(x_i) = q'_i(x_i) \qquad\qquad i = 1, \ \ldots, \ m - 1$$

$$q''_{i+1}(x_i) = q''_i(x_i) \qquad\qquad i = 1, \ \ldots, \ m - 1.$$

217

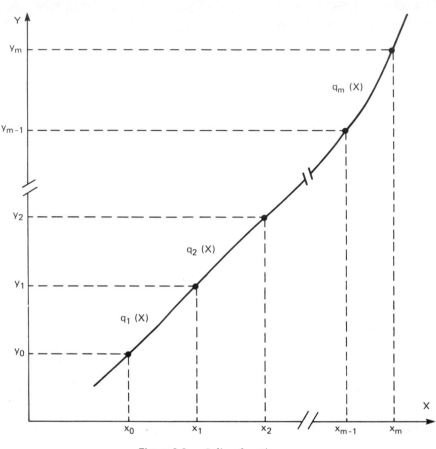

Figure 8-2   Spline functions.

In order to solve any system of algebraic equations, it is necessary for the number of equations to be exactly equal to the number of unknowns. At this point there are 4m unknowns and 4m - 2 constraint equations. This indicates that two more constraints are required. The usual choice for the two additional constraints is that:

$$q_1''(x_0) = 0 \qquad \text{and} \qquad q_m''(x_m) = 0$$

This choice leads to a spline representation known as the "natural cubic spline." Once the coefficients associated with a spline function are known, the piecewise polynomial function may be used to represent the data for purposes of interpolation, curve fitting, or surface fitting.

At first inspection the problem of determining the necessary coefficients might seem to be that of solving 4m equations for 4m unknowns.

218

Fortunately, however, with careful selection of the general form of the cubic polynomials, the complexity of the problem can be greatly simplified. If the individual cubic equations are selected to be:

$$q_i(x) = ty_i + \bar{t}y_{i-1} + \Delta x_i [(k_{i-1} - d_i)t\bar{t}^2 - (k_i - d_i)t^2\bar{t}]$$

$$i = 1, \ldots, m,$$

where:

$$\Delta x_i = x_i - x_{i-1}$$

$$t = \frac{x - x_{i-1}}{\Delta x_i}$$

$$\bar{t} = 1 - t$$

and

$$\Delta y_i = y_i - y_{i-1}$$

$$\frac{\Delta y_i}{\Delta x_i} = d_i,$$

then each of the $q_i(x)$ equations has only two constant coefficients to be determined. After the first $q_i(x)$ equation has been written each new equation adds only one new constant coefficient to be determined. For this formulation the value of $t = 0$ and $\bar{t} = 1$ for $x = x_{i-1}$ and $\bar{t} = 0$ and $t = 1$ for $x = x_i$. Thus all conditions except for second derivatives are automatically satisifed by this special formulation. The second derivative requirements give rise to the relationships:

$$k_{i-1}\Delta x_{i+1} + 2k_i(\Delta x_i + \Delta x_{i+1}) + k_{i+1}\Delta x_i = 3(d_i\Delta x_{i+1} + d_{i+1}\Delta x_i)$$

for internal points and

$$2k_0 + k_1 = 3d_1$$

and

$$k_{m-1} + 2k_m = 3d_m$$

for the two external points.

Thus the system of equations to be solved is a linear, tridiagonal system of the form:

$$
\begin{bmatrix}
2 & 1 & 0 & & 0 & \\
\Delta x_2 & 2(\Delta x_1 + \Delta x_2) & \Delta x_1 & & & \\
& \Delta x_3 & 2(\Delta x_2 + \Delta x_3) & \Delta x_2 & & \\
0 & & & & & \\
& & \Delta x_m & 2(\Delta x_{m-1} + \Delta x_m) & \Delta x_{m-1} & \\
& & & 1 & 2 &
\end{bmatrix}
\begin{bmatrix}
k_0 \\
k_1 \\
k_2 \\
\cdot \\
\cdot \\
k_m
\end{bmatrix}
= 3
\begin{bmatrix}
d_1 \\
d_1\Delta x_2 + d_2\Delta x_1 \\
d_2\Delta x_3 + d_3\Delta x_2 \\
\cdot \\
d_{m-1}\Delta x_m + d_m\Delta x_{m-1} \\
d_m
\end{bmatrix}
$$

For this equation system, the number of coefficients to be determined is equal to the number of data points. Thus the solution process is no more involved than that of fitting the m + 1 data points with an "mth-order" polynomial. It is not uncommon to find that a cubic spline function will do a better job of approximating the function than a polynomial of order m. It is worth mentioning that considerable variety exists in the use of spline functions because it is possible to choose other types of end constraints and because polynomials of order higher than three could be used.

8.4   Numerical Differentiation

Under certain circumstances it may be necessary to approximate the derivative of tabulated data. When this becomes necessary, the numerical formulas for approximating derivatives can be derived from a consideration of a Taylor's series expansion or by differentiation of the interpolation formulas developed in Section 8.2. Numerical differentiation of tabulated data can be a perilous process and should be avoided whenever possible. In order to understand the pitfalls associated with this process, it is necessary to consider two potential sources of difficulty. The first difficulty arises in the consideration of tabular information obtained by experimental means. The phenomenon of noise or experimental error is present to some degree in every experimental measurement. Thus the true signal shown in Figure 8-3(a) would be measured with noise as that shown in Figure 8-3(b). If this data is differentiated, the effect of the error is greatly amplified by the differentiation process. The result is shown in Figure 8-3(c). If, on the other hand, the data with noise is integrated as shown in Figure 8-3(d), the effect of the error is diminished. Thus numerical integration tends to be a far more stable process than that of numerical differentiation.

The second difficulty that arises in numerical differentiation is illustrated in Figure 8-4. Even though an interpolation polynomial may do an adequate job of describing the tabular function, its higher order derivatives may be totally different from that of the tabular function. This may be observed in Figure 8-4 by a comparison of the slopes and radii of curvature for the two curves.

To illustrate how numerical differentiation formulas can be obtained from Lagrangian interpolation formulas, a simple illustration can be used. Suppose that a second-order approximating polynomial of the form:

$$P(x) = C_0 + C_1(x - x_0) + C_2(x - x_0)(x - x_1)$$

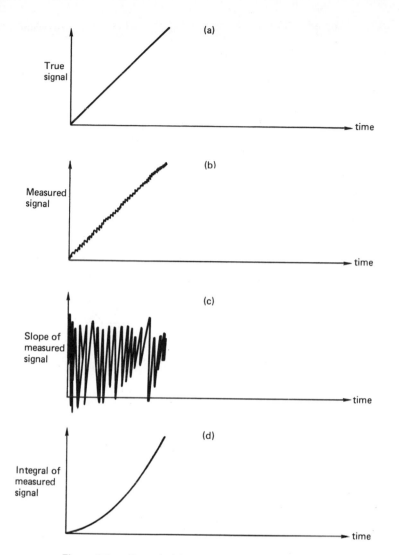

Figure 8-3   Numerical integration and differentiation.

is arranged so as to pass through three points

$$(x_0, y_0), \ (x_1, y_1), \ \text{and} \ (x_2, y_2).$$

By applying Lagrange's interpolation methodology, it can be shown that the constants will be:

$$C_0 = y_0$$

221

and

$$C_1 = \frac{y_1 - y_0}{h}$$

$$C_2 = \frac{y_2 - y_1 + y_0}{2h^2}.$$

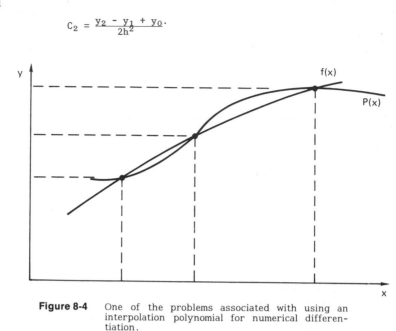

Figure 8-4   One of the problems associated with using an interpolation polynomial for numerical differentiation.

If the polynomial $P(x)$ is differentiated, the result will be:

$$P'(x) = C_1 + C_2(2x - x_0 - x_1).$$

If this expression is evaluated at $x = x_0$ using $(x_0 - x_1) = -h$, the result will be:

$$P'(x_0) = \frac{(-3y_0 + 4y_1 - y)}{2h}$$

This result is said to be a second order approximation for the derivative at the point $x = x_0$ in terms of forward points.

Whenever a numerical derivative is used, it is wise to consider the order of magnitude of the error associated with the approximation. To determine this quantity, a Taylor's expansion of the form:

$$f(x+\varepsilon) = f(x) + \frac{\varepsilon}{1} f'(x) + \frac{\varepsilon^2}{2!} f''(x) + \frac{\varepsilon^3}{3!} f'''(x) + \ldots.$$

will be used. If $x = x_0$ and $\varepsilon = h$ (one spacing), then $f(x_0 + h) = y_1$. Thus:

$$y_1 = y_0 + hD(y_0) + \frac{h^2}{2} D^2(y_0) + \frac{h^3}{6} D^3(y_0) + \ldots.$$

In like manner, if $\varepsilon = 2h$, the result will be:

$$y_2 = y_0 2hD(y_0) + 4\frac{h^2}{2}D^2(y_1) + 8\frac{h^3}{6}D^3(y_0) + \ldots.$$

If the second order derivative is eliminated from these two equations by multiplying the first by 4, and the second by -1 and then by adding, the result will be:

$$D(y_0) = \frac{4y_1 - y_2 - 3y_0}{2h} + \frac{h^2 y_0'''}{3}$$

This second order approximation to the first derivative is the same as found previously except that the error term $\frac{1}{3}h^2 y_0'''$ is present.

A similar procedure to those previously presented could be applied to obtain higher-order derivatives for forward, central, or backward forms. An illustration of the results for different types of interpolation polynomials is presented in Tables 8-2, 8-3, and 8-4.

8.5   Numerical Integration

Situations requiring the numerical evaluation of a definite integral often occur in engineering analysis. Quite frequently no closed-form solutions for these integrals can be found, or the complexity of the closed-form is such that numerical methods of evaluation are more practical. Numerical integration, also called numerical quadrature, is a stable process as illustrated in Figure 8-3. In contrast to numerical differentiation, numerical integration tends to minimize the influence of data errors on the final result. Numerical integration methods are based on approximating the definite integral as a sum of incremental areas. In general, the problem to be solved can be stated as that of finding the value of:

$$I = \int_a^b f(x)dx.$$

Numerical integration methods are classed according to whether the tabulated abscissa values are uniformly spaced or not. The Newton-Cotes formulas require equally spaced abscissa values, but the Gaussian formulas do not. Each of these two classes is now discussed.

Newton-Cotes Formulas

The simplest of all Newton-Cotes formulas comes from the trapezoidal method of integration. This method utilizes linear approximations to the function to be integrated. Lines are drawn between adjacent tabulated points (x_i, y_i) and (x_{i+1}, y_{i+1}) for $a \le x \le b$. Thus, if $x_0 = a$ and $x_n = b$, the integral will be represented as the sum of areas of the "n" trapezoids

Table 8-2 Derivative Approximation Formulas in Terms of Forward Positions

Derivative	2 Positions	3 Positions	4 Positions	5 Positions
y_0'	$\frac{1}{h}(y_1-y_0)$ $-\left(\frac{h}{2}y_0'\right)$	$\frac{1}{2h}(-y_2+4y_1-3y_0)$ $+\left(\frac{h^2}{3}y_0''''\right)$	$\frac{1}{6h}(2y_3-9y_2+18y_1-11y_0)$ $-\left(\frac{h^3}{4}y_0''''\right)$	$\frac{1}{12h}(-3y_4+16y_3-36y_2+48y_1-25y_0)$ $+\left(\frac{h^4}{5}y_0'''''\right)$
y_0''		$\frac{1}{h^2}(y_2-2y_1+y_0)$ $-(hy_0''')$	$\frac{1}{h^2}(-y_3+4y_2-5y_1+2y_0)$ $+\frac{11h^2}{12}y_0''''$	$\frac{1}{12h^2}(11y_4-56y_3+114y_2-104y_1+35y_0)$ $+\left(\frac{5h^3}{6}y_0'''''\right)$
y_0'''			$\frac{1}{h^3}(y_3-3y_2+3y_1-y_0)$ $-\left(\frac{3h}{2}y_0''''\right)$	$\frac{1}{2h^3}(-3y_4+14y_3-24y_2+18y_1-5y_0)$ $+\left(\frac{21h^2}{12}y_0'''''\right)$

Note: The error term is shown in parentheses following each formula.

Table 8-3   Derivative Approximations in Terms of Central Positions

Derivative	3 Positions	5 Positions	7 Positions
y'_0	$\frac{1}{2h}(y_1 - y_{-1})$ $-(\frac{h^2}{6}y''')$	$\frac{1}{12h}(-y_2 + 8y_1 - 8y_{-1} + y_{-2})$ $+(\frac{h^4}{30}y_0^v)$	$\frac{1}{60h}(y_3 - 9y_2 + 45y_1 - 45y_{-1} + 9y_{-2} - y_{-3})$ $-(\frac{h^6}{140}y_0^{vii})$
y''_0	$\frac{1}{h^2}(y_1 - 2y_0 + y_{-1})$ $-(\frac{h^2}{12}y'''')$	$\frac{1}{12h^2}(-y_2 + 16y_1 - 30y_0 + 16y_{-1} - y_{-2})$ $-(\frac{h^4}{90}y_0^{vi})$	$\frac{1}{180h^2}(2y_3 - 27y_2 + 270y_1 - 490y_0 + 270y_{-1} - 27y_{-2} + 2y_{-3})$ $-(\frac{h^6}{560}y_0^{viii})$
y'''_0		$\frac{1}{2h^3}(y_2 - 2y_1 + 2y_{-1} - y_{-2})$ $-(\frac{h^2}{4}y_0^v)$	$\frac{1}{8h^3}(-y_3 + 8y_2 - 13y_1 + 13y_{-1} - 8y_{-2} + y_{-3})$ $+(\frac{7h^4}{120}y_0^{vii})$

Note: The error term is shown in parentheses following each formula.

Table 8-4   Derivative Approximations in Terms of Backward Positions

Derivative	2 Positions	3 Positions	4 Positions	5 Positions
y_0'	$\frac{1}{h}(y_0 - y_{-1})$ $+ (\frac{h}{2}y_0'')$	$\frac{1}{2h}(3y_0 - 4y_{-1} + y_{-2})$ $+ (\frac{h^2}{3}y_0''')$	$\frac{1}{6h}(11y_0 - 18y_{-1} + 9y_{-2} - 2y_{-3})$ $+ (\frac{h^3}{4}y_0'''')$	$\frac{1}{12h}(25y_0 - 48y_{-1} + 36y_{-2} - 16y_{-3} + 3y_{-4})$ $+ (\frac{h^4}{5}y_0''''')$
y_0''		$\frac{1}{h^2}(y_0 - 2y_{-1} + y_{-2})$ $+ (hy_0''')$	$\frac{1}{h^2}(2y_0 - 5y_{-1} + 4y_{-2} - y_{-3})$ $+ (\frac{11h^2}{12}y_0'''')$	$\frac{1}{h^2}(35y_0 - 104y_{-1} + 114y_{-2} - 56y_{-3} + 11y_{-4})$ $+ (\frac{5h^3}{6}y_0''''')$
y_0'''			$\frac{1}{h^3}(y_0 - 3y_{-1} + 3y_{-2} - y_{-3})$ $+ (\frac{3h}{2}y_0'''')$	$\frac{1}{2h^3}(5y_0 - 18y_{-1} + 24y_{-2} - 14y_{-3} + 3y_{-4})$ $+ (\frac{7h^2}{4}y_0''''')$

Note: The error term is shown in parentheses following each formula.

each having a height of "h". The integral value written in terms of the tabular ordinates will be:

$$I = \int_a^b f(x)dx = \frac{h}{2} [y_0 + 2y_1 + 2y_2 + \ldots, +2y_{n-1} + y_n].$$

This trapezoidal method is illustrated graphically in Figure 8-5. Clearly, as the spacing between tabular points gets small, the accuracy of this approximation improves.

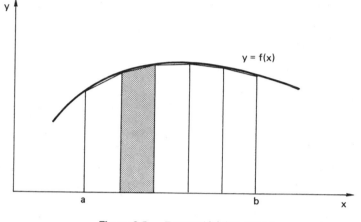

Figure 8-5 Trapezoidal integration.

The concept of using linear approximations of the function to determine the area of subintervals leads quite naturally to the idea that one might use a higher-order curve such as a parabola. Of course, a parabola requires three adjacent tabular points rather than two. As a result the numerical formula for the integral differs from that found for trapezoidal segments. This new formula, known as Simpson's rule, is:

$$I = \int_a^b f(x)dx = \frac{h}{3} [y_0 + 4y_1 + 2y_2 + 4y_3 + \ldots, +4y_{n-1} + y_n].$$

This equation must be applied to an even number of intervals. The fact that formulas can be found for polynomials of first and second order leads to the desire to generalize for higher-order polynomials. A detailed mathematical treatment of this situation is presented by Ralston [9]. The result is a general form:

$$\int_a^b f(x)dx = C_0 h \sum_{i=0}^{h} w_i f_i + C_1 h^{k+1} f^{(k)}(x^*),$$

227

Table 8-5 Coefficients for the Newton-Cotes Formulas

k	C_0	w_0	w_1	w_2	w_3	w_4	w_5	w_6	w_7	w_8	C_1
1	$\frac{1}{2}$	1	1								$-\frac{1}{12}$
2	$\frac{1}{3}$	1	4	1							$-\frac{1}{90}$
3	$\frac{3}{8}$	1	3	3	1						$-3/80$
4	$\frac{2}{45}$	7	32	12	32	7					$-8/945$
5	$\frac{5}{288}$	19	75	50	50	75	19				$-275/12096$
6	$\frac{1}{140}$	41	216	27	272	27	216	41			$-9/1400$
7	$\frac{7}{17280}$	751	3577	1323	2989	2989	1323	3577	751		$-8183/518400$
8	$\frac{4}{14175}$	989	5888	-928	10946	-4540	10946	-928	5888	989	$-2368/467775$

where n is the number of strips, k is the order of the polynomial used, x* is some point in the interval [a,b], $f^{(k)}(x*)$ is the kth derivative of the function f(x) evaluated at x*, and h is the width of the data intervals. The coefficients C_0, C_1, and w_i are listed in Table 8-5. Each row of Table 8-5 represents a cycle of k-strips involving k + 1 pivotal points to achieve a kth order polynomial. To apply this information to a problem of more than one cycle (say k = 2 and n = 6), the coefficients should be added end to end so that the last weight values overlap. Thus:

	w_0	w_1	w_2	w_3	w_4	w_5	w_6
	1	4	1				
			1	4	1		
					1	4	1
sum	1	4	2	4	2	4	1

This gives:

$$\int_a^b f(x)dx = \frac{h}{3}\,[y_0+4y_1+2y_2+4y_3+2y_4+4y_5+y_6] + C_1 h^3 f^{(3)}(x*),$$

which agrees with the result previously presented. The final term in the Newton-Cotes formula:

$$C_1 h^{k+1} f^{(k)}(x*),$$

is a term that indicates the order of magnitude of the error associated with the approximation. Clearly, as h gets small, h^{k+1} gets even smaller. This trend should be used with care, however, because the error also depends on $f^{(k)}(x*)$. There do exist some functions whose higher order derivatives become very large. For these functions the error may not be reduced by increasing the value of k used.

Gaussian Quadrature
All Newton-Cotes integration formulas involve the use of equally spaced abscissa points. If this restriction is lifted, it is possible to choose the spacing of the data points in such a way that the error in the approximation is reduced. This is the basis of Gaussian quadrature.

In the formula:

$$\int_a^b f(x)dx = \sum_{i=0}^{n} w_i f(x_i) + E,$$

229

both w_i and x_i are treated as unknowns to be determined. Thus the total number of unknowns to be determined will be $2(n + 1)$. A polynomial of order $2n + 1$ requires $2n + 2$ conditions to specify uniquely its form. Thus we will approximate the integral using a polynomial of order $2n + 1$ and require that the approximation have zero error for all polynomials of degree less than or equal to this value. This process will give $2n + 2$ equations in $2n + 2$ unknowns. In general, the resulting equations will be linear in w_i but nonlinear in the x_i values. If the limits of integration are $[-1, 1]$, the x_i values will be the $(n + 1)$ roots of the Legendre polynomial $p_{n+1}(x) = 0$. Once the x_i values are known, the w_i values can be found by linear methods. To illustrate the use of this information, let us consider a problem in which $n = 1$. The requirement that

$$\int_{-1}^{1} f(x)\,dx = \sum_{i=0}^{1} w_i f(x_i)$$

be satisfied for all polynomials of degree $2n + 1$ and less means that if one uses $f(x) = 1, x, x^2$, or x^3, the equality should hold with no error. This gives the following equations:

$$w_0 + w_1 \quad = \int_{-1}^{+1} dx \quad = 2$$

$$w_0 x_0 + w_1 x_1 \quad = \int_{-1}^{+1} x\,dx \quad = 0$$

$$w_0 x_0^2 + w_1 x_1^2 \quad = \int_{-1}^{+1} x^2\,dx = 2/3$$

$$w_0 x_0^3 + w_1 x_1^3 \quad = \int_{-1}^{+1} x^3\,dx = 0$$

This system of four equations and four unknowns can be solved even though it is nonlinear. To shortcut the nonlinear difficulty, the Legendre polynomial,

$$p_2(x) = [-1 + 3x^2],$$

can be used. This polynomial will have roots $x_1 = -\sqrt{3}/3$ and $x_2 = +\sqrt{3}/3$. Using these values and any two of the constraint equations shown above, one can find $w_0 = 1.$ and $w_1 = 1.$ Thus the integral can be expressed as:

$$\int_{-1}^{+1} f(x)dx = f\left(\frac{-\sqrt{3}}{3}\right) + f\left(\frac{+\sqrt{3}}{3}\right).$$

The restriction that the interval of integration must be $[-1, 1]$ is not as confining as one might think. Clearly, a change of variable:

$$x = \frac{2Z - (a + b)}{b - a}$$

$$F(Z) = f(x)$$

will convert a general integral to the required form:

$$I = \int_{a}^{b} F(Z)dZ = \frac{b - a}{2} \int_{-1}^{+1} f(x)dx.$$

It is worthwhile to mention that Gaussian formulas can be applied to equal-sized subintervals and the values added to find the area of a total interval.

Problems

8.1 Uniformly spaced values from a set of thermocouple tables are presented below. Perform interpolation to find the value at T = 55°F from this data using Lagrange interpolation.

T°F	MV
0	-0.670
20	-0.254
40	0.171
60	0.609
80	1.057
100	1.517

8.2 Solve Problem 8.1 by using the method of divided differences.

8.3 Solve Problem 8.1 using a cubic spline function.

8.4 Elliptic integrals† arise when using the exact form of the Euler-Bernoulli equation for the large deflection of beams. The Elliptic integral of the first type is defined as:

$$F(k,\phi) = \int_{0}^{\phi} \sqrt{(1 - k^2\sin^2\theta)} \, d\theta$$

---

†Abramowitz, M. and Stegun, I. A. Handbook of Mathematical Functions, National Bureau of Standards Applied Mathematics Series 55. U.S. Government Printing Office, Washington, D.C., 1964.

where $0 \leq k \leq 1.0$. Prepare a FORTRAN function subprogram ELLIP(XK,PHI) that will evaluate the Elliptic integral for given values of the arguments. Test the accuracy of your subprogram using:

$$F(0.5, \pi/6) = 0.52942863.$$

8.5 The Sievert integral† is often used in engineering practice because it is related to the error function and to the integral of the Bessel function. The Sievert integral is defined as:

$$S(x,\theta) = \int_0^\theta e^{-x \sec \phi} \, d\phi.$$

Prepare a FORTRAN function subprogram SIEV(X,THETA) that will evaluate the Sievert integral using Simpson's rule. Test the accuracy of your subprogram using:

$$S(1.0, \pi/3) = 0.307694.$$

8.6 Frequently in engineering problem solving, data is available in graphical form only. In order to use this information in a computer-aided design process, it is necessary to be able to describe this information analytically. For the information displayed in the curves below, the nondimensional load-vs-deflection curve for $\alpha = 90°$

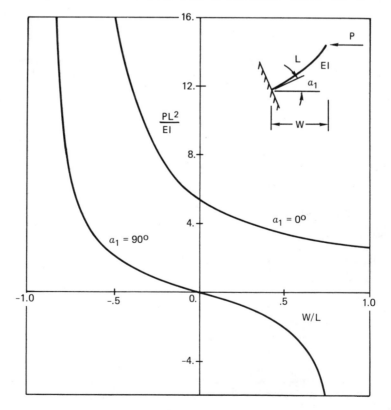

exhibits antisymmetry. Read values on this curve corresponding to:

$$W/L = -0.1, \ -0.2, \ -0.3, \ -0.4, \ -0.5, \ -0.6, \ -0.7, \ \text{and} \ -0.8.$$

Use a least squares to fit to describe this curve by a polynomial of order 3, 5, and 7. Which to you think is best? Why?

8.7 For the data gathered in Problem 8.6, find an approximation for the first derivative of the data. Compare this result to the derivatives of the least squares polynomials.

8.8 For the curve $\alpha = 0°$ in Problem 8.6, read data values from:

$$-0.5 \le W/L \le 1.0$$

and fit this information with a cubic spline function. Find the area under the curve by numerical integration.

8.9 The data in the accompanying chart† shows the average height for male subjects between the ages of 4 and 17 years. Von Bertalanffy†† has proposed that a good mathematical model for growth data is:

$$y = a(1 - be^{-kt}),$$

where a, b, and k are constants to be determined. Using the data provided and the proposed growth curve, find the best values of a, b, and k using a least squares fit.

Age in Years	Average Height (inches)
4	40.9
5	43.9
6	46.1
7	48.2
8	50.4
9	52.4
10	54.3
11	56.2
12	58.2
13	60.5
14	63.0
15	65.6
16	67.3
17	68.2

8.10 For the data in Problem 8.9, find and plot the growth rate (the time rate of change of growth). For what age is this maximum?

---

†Martin, W. E. Basic Body Measurements of School Age Children. U. S. Dept. of Health, Education, and Welfare, Office of Education, Washington, D.C., June 1953.

††von Bertalanffy, L. "Quantitative laws in metabolism and growth." Quarterly Review of Biology, 32 (1957): 217-31.

References

1. Abramowitz, M. and Stegun, I. A., eds. Handbook of Mathematical Functions with Formulas, Graphs and Mathematical Tables. U.S. Government Printing Office, Washington, D.C., 1964.

2. Ahlberg, H. J., Nilson, E. N., and Walsh, J. L. The Theory of Splines and Their Application. New York: Academic Press, 1967.

3. Bickley, W. G. "Formulae for Numerical Differentiation." Mathematical Gazette, 25 (1941): 19-27.

4. Grove, W. E. Brief Numerical Methods. Englewood Cliffs, N.J.: Prentice-Hall, Inc., 1966.

5. Ketter, R. L., and Prawel, S. P., Jr. Modern Methods of Engineering Computation. New York: McGraw-Hill Book Co., 1969.

6. LaFara, R. L. Computer Methods for Science and Engineering. Rochelle Park, N.J.: Hayden Book Co., 1973.

7. McCalla, T. R. Introduction to Numerical Methods and FORTRAN Programming. New York: John Wiley & Sons, Inc., 1967.

8. Pall, G. A. Introduction to Scientific Computing. New York: Meredith Corp., 1971.

9. Ralston, A. A First Course in Numerical Analysis. New York: McGraw-Hill Book Co., 1965.

10. Williams, P. W. Numerical Computation. New York: Harper & Row Publishers, Inc., 1972.

APPENDIX A
COMPUTER SOFTWARE FOR
ENGINEERING APPLICATIONS

A wide selection of prewritten computer software is presently available to assist the engineering problem solver. This appendix contains a list of about 100 such programs. This particular group has been selected because it contains programs that are generally useful and are fairly well known. Although the author has not had first-hand experience with all of these programs, they are all available from reputable sources. Many of these programs are proprietary and are available only through subscription arrangements. In most cases these programs and subroutine packages have undergone considerable development and improvement, and most can be implemented on a variety of different computer systems.

NAME	APPLICATION	AVAILABLE
ADLPIPE	A special purpose program for the static and dynamic analysis of complex piping systems. Developed by the Arthur D. Little Co.	McDonnell Douglass Automation Co. Box 516 St. Louis, MO
AFL	Static force analysis of four-bar linkage systems with a hydraulic cylinder force input.	Structural Dynamics Research Corp. Cincinnati, OH 45227
ANSYS	A general-purpose program for the solution of a large class of engineering analysis problems including static and dynamic analysis, elastic and plastic analysis, fluid flow and transient heat transfer analysis.	Structural Dynamics Research Corp. Cincinnati, OH 45227
APTMILL	A two-dimensional contouring, three-dimensional positioning language designed for milling machines and machining center operations.	Structural Dynamics Research Corp. Cincinnati, OH 45227

NAME	APPLICATION	AVAILABLE
APTURN	Computer program for preparing tapes for numerically controlled lathes.	Structural Dynamics Research Corp. Cincinnati, OH 45227
ASTRA	A general purpose finite element code.	Boeing Computer Services, Inc. P. O. Box 24346 Seattle, WA 98124
AUTOFLEX	Pipe flexibility program.	United Computer Systems, Inc. 4544 Post Oak Place Houston, TX 77027
BEAMSTRESS	A computer program that uses a finite element representation to determine the section properties and stresses in an arbitrary homogeneous cross section of a straight beam.	United Computing Systems, Inc. 4544 Post Oak Place Houston, TX 77027
BRAD	Bearing stiffness and deflection calculations for tapered roller bearings, plain roller bearings, spherical roller bearings and angular contact ball bearings.	Structural Dynamics Research Corp. Cincinnati, OH 45227
CAMP	Digital analog simulation program that yields real time response of a described system.	Structural Dynamics Research Corp. Cincinnati, OH 45227
CAMPAC	Synthesis, analysis, and design of cams.	Prof. D. Tesar University of Florida Gainesville, FL 32601
CIRP2	Analysis of circular plates based on classical engineering plate theory.	United Computing Systems, Inc. 4544 Post Oak Place Houston, TX 77027
COGO	A program for solving coordinate geometry problems commonly encountered by the civil engineer.	United Computing Systems, Inc. 4544 Post Oak Place Houston, TX 77027
COMMEND I	A generalized mechanical design system incorporating linkage, cam, gear, spring, shaft, and timing-belt design programs, together with related routines for N/C machining	IBM Systems Development Division Development Laboratory Rochester, MN 55901
CSMP	Continuous Systems Modeling Program. Non-linear transient response of continuous systems represented by algebraic equations, differential equations, and various functional blocks.	IBM 112 E. Post Road White Plains, NY 10601
DAD	Reduces, displays, and analytically represents test data obtained from transfer function analyzer of digital Fourier analyzer with various A/D converters.	Structural Dynamics Research Corp. Cincinnati, OH 45227

236

NAME	APPLICATION	AVAILABLE
DAD II	Performs general time series operations on data which are evenly spaced in the time or frequency domain.	Structural Dynamics Research Corp. Cincinnati, OH 45227
DAGS	Dynamic analysis of general structures.	Structural Dynamics Research Corp. Cincinnati, OH 45227
DALF1	Dynamic analysis of linear frames.	Mr. Larry M. Bryant Dept. of Civil Eng. University of Texas Austin, TX 78712
DANF1	Dynamic analysis of nonlinear frames.	Mr. Larry M. Bryant Dept. of Civil Eng. University of Texas Austin, TX 78712
DDAM	Dynamic design/analysis of structures composed of beams and/or springs.	Computer Sciences Corp. INFONET Division 650 N. Sepulveda El Segundo, CA 90245
DISK	Calculation of the radial and tangential stresses in rotating disks of variable thickness.	Structural Dynamics Research Corp. Cincinnati, OH 45227
DKINAL	Dynamic analysis of machinery.	Prof. B. Paul Dept. of Mech. Eng. University of Penn. Philadelphia, PA 19174
DRAIN 2D	Dynamic response of inelastic, two-dimensional structures of arbitrary configuration resulting from earthquake type ground motions.	NISEE University of Calif. Berkeley, CA 94729
DRAM	Interactive simulation of two-dimensional dynamic mechanical systems that are open loop or closed loop and are single or multiple degree of freedom. Developed by M. Chace, Univ. of Mich.	Structural Dynamics Research Corp. 5729 Dragon Way Cincinnati, OH 45227
DRPL	Three-dimensional dynamic response of mechanisms.	Prof. Carson University of Iowa Iowa City, IA 52242
DYAD	Dynamic analysis of planar mechanisms.	Prof. D. Tesar University of Florida Gainesville, FL 32601
DYNAFLEX	A program designed to analyze piping systems subjected to dynamic loadings as well as static loadings.	United Computing Systems, Inc. Houston, TX
DYNAPLAS	Dynamic, large deflection, elastic-plastic analysis of stiffened shells of revolution.	United Computing Systems, Inc. Houston, TX

NAME	APPLICATION	AVAILABLE
DYNATIER	Modes and frequencies, and response of structures to piecewise linear forcing for rectangular space frames.	Prof. W. Weaver Dept. of Civil Eng'e. Stanford University Stanford, CA 94305
DYSIN	Group of Burmester theory programs based on four-bars, six-bars, and geared five-bar linkages.	Prof. D. Tesar University of Fla. Gainsville, FL 32601
EARTH	A program for calculating subdivision and roadway earthwork quantities.	United Computing Systems, Inc. 4544 Post Oak Place Houston, TX 77027
EISPAK	A special package of subroutines for eigenanalysis.	Argonne National Labs 9700 S. Cass Ave. Argonne, IL 60439
ETC	Calculates torsional constant of standard cross sections.	Structural Dynamics Research Corp. Cincinnati, OH 45227
FIVEPOS	Five-position Burmester theory. Synthesis of function, path, and motion generating four-bar linkages.	Prof. G. N. Sandor University of Fla. Gainesville, FL 32601
FLOORR	A program that designs building floor systems and supporting columns.	United Computing Systems, Inc. 4544 Post Oak Place Houston, TX 77027
FLOORT	A computer program for designing building floors of composite construction including cover plates and material quantity estimates.	United Computing Systems, Inc. 4544 Post Oak Place Houston, TX 77027
FMA	Modes and frequencies of space frames.	COSMIC University of Georgia Athens, GA 30601
FORBAR	Kinematic and dynamic analysis of four-bar linkage systems. Output of program includes position, velocity, and acceleration of each link, inertia forces, and pin reactions.	Structural Dynamics Research Corp. Cincinnati, OH 45227
FRAME	Static analysis, modes and frequencies, and transient displacement by mode superposition and direct integration of linearly elastic frames.	Nippon Univac Co. 17-51 Akaska 2-Chome Minato-ku Tokyo 107 JAPAN
FUNPAK	Special function package containing Bessel functions, and elliptic and exponential integrals.	Argonne National Labs 9700 S. Cass Ave. Argonne, IL 60439
GPSS	A simulation programming language used to build computer models for discrete - event simulation.	IBM 112 E. Post Road White Plains, NY 10601

NAME	APPLICATION	AVAILABLE
GRID	Analysis of grillages with up to 11 girders and 50 stiffeners.	United Computing Systems, Inc. 4544 Post Oak Place Houston, TX 77027
IMP	Integrated mechanisms package. Analysis of two or three dimensional closed-loop rigid link mechanisms with all types of joints, linear springs, viscous dampers, mass, and gravity effects.	Prof. J. J. Uicker University of Wisconsin Madison, WI 53706
INELASTIER	Static and transient response of plane frames or of rectangular buildings using tier building models.	Prof. William Weaver Dept. of Civil Eng'g. Stanford University Stanford, CA 94305
IOWA CADET	A battery of more than 150 subprograms to perform mathematical and statistical tasks upon which can be superimposed optimization strategies.	Dr. C. R. Mischke Iowa State University Ames, IA 50010
ISOCOM	Dynamic response analysis of resiliently mounted components (e.g., engines, machines and bases).	Structural Dynamics Research Corp. Cincinnati, OH 45227
JET 3	A computer program to calculate the large elastic-plastic dynamically-induced deformations of free or restrained, partial and/or complete structural rings.	Aeroelastic and Structural Research Lab MIT Cambridge, MA 02139
KINAL	Kinematic analysis of planar, multiloop single-degree-of-freedom mechanisms.	Prof. B. Paul Mech. Eng. Dept. University of Penn. Philadelphia, PA 19174
LAGS	A generalized elastic-plastic design analysis program for two and three-dimensional beam frameworks.	Structural Dynamics Research Corp. Cincinnati, OH 45227
LINK-1	A program to compute the static stability and dynamic responses for beams, long rectangular plates, thin-walled beams, and axisymmetric cylindrical shells.	United Computing Systems, Inc. 4544 Post Oak Place Houston, TX 77027
LINK-3	A rotating shaft analysis program to solve problems of critical speeds, unbalanced response, influence coefficients, and dynamic stability.	United Computing Systems Inc. 4544 Post Oak Place Houston, TX 77027
MARC	A general-purpose finite element program for the analysis of structures. Developed by the MARC Analysis Corporation, Providence, RI.	Control Data Corp. P. O. Box 0 Minneapolis, MN 55440
MECA	Calculates cutting forces, power, and cutting stiffness coefficients for a large variety of metal machining operations.	Structural Dynamics Research Corp. Cincinnati, OH 45227

Appendix A (continued)

NAME	APPLICATION	AVAILABLE
MECOL	Pipe flexibility analysis program.	United Computing Systems, Inc. 4544 Post Oak Place Houston, TX 77027
MODAL	A generalized computer program for analyzing the frequency or transient response of complex systems.	Structural Dynamics Research Corp. Cincinnati, OH 45227
NASTRAN	Static and dynamic structural analysis by the finite element approach. Buckling analysis, compressible fluids in tanks and cavities, and heat transfer analysis.	COSMIC University of Georgia Athens, GA 30601
NUPIPE	A special purpose program for the analysis of linearly elastic piping systems.	Nuclear Services Corp. 1700 Dill Avenue Campbell, CA 95008
OBL	Optimization bearing locations for one, two, or three bearings on a circular shaft in order to minimize the static deflection at any point on the shaft.	Structural Dynamics Research Corp. Cincinnati, OH 45227
OPTISEP	Designers' optimization subroutines.	Prof. J. N. Siddall College of Engineering McMaster University Hamilton, Ontario, CANADA
PCA	A series of programs to develop design solutions for airport pavement, bridges, floor systems, flat plates, and frames.	United Computing Systems, Inc. 4544 Post Oak Place Houston, TX 77027
PIFA	Pipe flexibility analysis program.	United Computing Systems, Inc. 4544 Post Oak Place Houston, TX 77027
PIPDYN	A finite element piping program based on the displacement method. Developed by the Franklin Institute Research Laboratories.	Utility Network of America 7540 LBJ Freeway Suite 830 Dallas, TX 75240
PIPE	Extended pipe flexibility analysis program.	United Computing Systems, Inc. 4544 Post Oak Place Houston, TX 77027
PIPERUP	A special purpose program for the analysis of piping systems.	Nuclear Service Corp. 1700 Dill Avenue Campbell, CA 95008
PIPESD	A finite element program for calculating the dynamic response of piping systems using modal response spectra.	Control Data Corp. P.O. Box O Minneapolis, MN 55440

NAME	APPLICATION	AVAILABLE
PREBEST	Calculates beam cross-section properties including location of the centroid, principal axes, area moments of inertia, cross section area, weight, and weight moments of inertia.	Structural Dynamics Research Corp. Cincinnati, OH 45227
PRINST	Analyzes experimental data from 45° and 60° strain rosettes and calculates principle stresses and strains.	Structural Dyanmics Research Corp. Cincinnati, OH 45227
RECPLAT	The program determines the displacement, slope, moments, natural frequencies, mode shapes, and buckling loads of rectangular plates.	United Computing Systems, Inc. 4544 Post Oak Place Houston, TX 77027
ROARKS	More than 1800 formulas from Professor Roark's text, Formulas for Stress and Strain.	United Computing Systems, Inc. 4544 Post Oak Place Houston, TX 77027
SABBA	System analysis via the building-block approach. Dynamic analysis of mechanical, structural, and electrical systems.	Structural Dynamics Research Corp. Cincinnati, OH 45227
SAGS	Static analysis of general structures.	Structural Dynamics Research Corp. Cincinnati, OH 45227
SAMMSOR II	Stiffness and mass matrices for shells of revolution.	United Computing Systems, Inc. 4544 Post Oak Place Houston, TX 77027
SAMMSOR III	Stiffness and mass matrices for ring-stiffened shells of revolution.	United Computing Systems, Inc. 4544 Post Oak Place Houston, TX 77027
SAP IV	General purpose program for finite element analysis of linearly elastic structural systems.	Earthquake Engineering Research Center University of Calif. Berkeley, CA 94720
SASA	Comprehensive calculation of all section properties of an arbitrarily shaped cross-section using finite element methods.	Structural Dynamics Research Corp. Cincinnati, OH 45227
SEAL-SHELL	A computer program that determines stresses, loads, and deflections of a thin shell of revolution with axisymmetric end loads and uniform pressure.	United Computing Systems, Inc. 4544 Post Oak Place Houston, TX 77027
SECS	A program for engineers and surveyors for solving subdivision geometry and plot problems.	United Computing Systems, Inc. 4544 Post Oak Place Houston, TX 77027

NAME	APPLICATION	AVAILABLE
SLIDER	Static and dynamic analysis of slider crank systems.	Structural Dynamics Research Corp. Cincinnati, OH 45227
SOILTIER	Soil-foundation-structure interaction during earthquakes. Analysis of modes and frequencies and of transient response of buildings.	Prof. W. Weaver Dept. of Civil Eng'g. Stanford University Stanford, CA 94305
SPIN	Static and dynamic in-plane bending analysis of beams and rotating shafts on elastic foundations.	Structural Dynamics Research Corp. Cincinnati, OH 45227
SPOTS	Calculation of cross-section properties of symmetric and unsymmetric thin walled cross sections.	Structural Dynamics Research Corp. Cincinnati, OH 45227
STARDYNE	A general-purpose finite element program for static and dynamic analysis of elastic structures.	Control Data Corp. P. O. Box O Minneapolis, MN 55440
STRUDL II	Static and dynamic analysis of structures. Dynamic analysis of trusses, frames, and structures made of two-dimensional membranes, bending elements, and three-dimensional elements.	McDonnell Douglass P. O. Box 516 St. Louis, MO 63100
STRU-PAK	Modes and frequencies of frame and truss structures.	Control Data Corp. P. O. Box O Minneapolis, MN 55440
SUPERB	A general-purpose finite element computer program.	Structural Dynamics Research corp. Cincinnati, OH 45227
TASS	Static and dynamic torsional analysis of single branched geared shaft systems.	Structural Dynamics Research Corp. Cincinnati, OH 45227
TEPC-2D	Two-dimensional finite element computer program for thermal-elastic-plastic-creep analysis.	General Atomic Company P. O. Box 81608 San Diego, CA 92138
TORK	Calculates torsional constant of arbitrarily shaped single- or multi-celled thin-walled cross section.	Structural Dynamics Research Corp. Cincinnati, OH 45227
TRAMFO	General nonlinear transient system featuring a sophisticated library of analog-type routines, combined with the efficient flexibility of FORTRAN.	Structural Dynamics Research Corp. Cincinnati, OH 45227
TRAP	Determination of the time response of a system due to an arbitrary force signal, using convolution integral.	Structural Dynamics Research Corp. Cincinnati, OH 45227

Appendix A (continued)

NAME	APPLICATION	AVAILABLE
TRIFLEX	Extended pipe flexibility analysis program.	United Computing Systems, Inc. 4544 Post Oak Pl. Houston, TX 77027
UNIVALVE II	A code to calculate the large deflection dynamic response of beams, rings, plates, and cylinders.	Sandia Laboratories SC-RR-61-303 Albuquerque, NM 87115
VISCOSUPERB	Finite element computer program for predicting static and dynamic behavior of viscoelastic materials.	Structural Dynamics Research Corp. Cincinnati, OH 45227
WECAN	A large general-purpose finite element program for linear and nonlinear structural analysis.	Westinghouse Electric Power Systems Co. Advanced Systems, Tech. 700 Braddock Ave. East Pittsburgh, PA 15112
XTABS	Three-dimensional analysis of building systems. Linear analysis of frame and shear wall buildings subjected to static and earthquake loading.	NISEE/Computer Appl. University of Calif. Berkeley, CA 94729

APPENDIX B
GUIDELINES FOR
PREPARING ENGINEERING
COMPUTER PROGRAMS

In preparing a computer program to solve a given engineering task, certain guidelines are worth keeping in mind:

1. Insofar as is possible, programs and subroutines should be self-documenting. A generous use of comment statements at the beginning of a program can prove helpful in explaining such things as:

   (a) the purpose of the program;
   (b) the inputs and outputs and their units;
   (c) the method used and its limitations or special requirements;
   (d) any appropriate references to the solution method;
   (e) the name of the program writer; and
   (f) the date that the program was prepared.

2. Sufficient documentation on the output pages should be provided in order to have the output make sense without the need to refer to the program. This should include:

   (a) a general header explaining what the output is;
   (b) a listing of all input values used including the units of these; and
   (c) an orderly arrangement of the output quantities including the units of these.

3. Whenever possible, the output of a program should be confined to 8½" x 11" size to facilitate its storage.

4. Programs and subroutines should have checks for bad input. This should include appropriate diagnostic messages concerning the errors detected.

5. Test data for which there is a known solution should be provided that will test all loops and options in a routine. This data and its output should be kept on file along with a listing of the program to provide a means to recheck a program whenever modifications are made.

APPENDIX C
CONVERSION FACTORS

These factors have been reproduced from NASA SP-7012, pp. 15-20. The first two digits of each numerical entry represent a power of 10. An asterisk follows each number that expresses an exact definition.

LISTING BY PHYSICAL QUANTITY

ACCELERATION

foot/second²	meter/second²	−01 3.048*
free fall, standard	meter/second²	+00 9.806 65*
gal (galileo)	meter/second²	−02 1.00*
inch/second²	meter/second²	−02 2.54*

AREA

acre	meter²	+03 4.046 856 422 4*
are	meter²	+02 1.00*
barn	meter²	−28 1.00*
circular mil	meter²	−10 5.067 074 8
foot²	meter²	−02 9.290 304*
hectare	meter²	+04 1.00*
inch²	meter²	−04 6.4516*
mile² (U.S. statute)	meter²	+06 2.589 988 110 336*
section	meter²	+06 2.589 988 110 336*
township	meter²	+07 9.323 957 2
yard²	meter²	−01 8.361 273 6*

DENSITY

gram/centimeter³	kilogram/meter³	+03 1.00*
lbm/inch³	kilogram/meter³	+04 2.767 990 5
lbm/foot³	kilogram/meter³	+01 1.601 846 3
slug/foot³	kilogram/meter³	+02 5.153 79

ENERGY

To convert from	to	multiply by
British thermal unit:		
(IST before 1956)	joule	+03 1.055 04
(IST after 1956)	joule	+03 1.055 056
British thermal unit (mean)	joule	+03 1.055 87
British thermal unit (thermochemical)	joule	+03 1.054 350
British thermal unit (39° F)	joule	+03 1.059 67
British thermal unit (60° F)	joule	+03 1.054 68
calorie (International Steam Table)	joule	+00 4.1868
calorie (mean)	joule	+00 4.190 02
calorie (thermochemical)	joule	+00 4.184*
calorie (15° C)	joule	+00 4.185 80
calorie (20° C)	joule	+00 4.181 90
calorie (kilogram, International Steam Table)	joule	+03 4.1868
calorie (kilogram, mean)	joule	+03 4.190 02
calorie (kilogram, thermochemical)	joule	+03 4.184*
electron volt	joule	−19 1.602 191 7
erg	joule	−07 1.00*
foot lbf	joule	+00 1.355 817 9
foot poundal	joule	−02 4.214 011 0
joule (international of 1948)	joule	+00 1.000 165
kilocalorie (International Steam Table)	joule	+03 4.1868
kilocalorie (mean)	joule	+03 4.190 02
kilocalorie (thermochemical)	joule	+03 4.184*
kilowatt hour	joule	+06 3.60*
kilowatt hour (international of 1948)	joule	+06 3.600 59
ton (nuclear equivalent of TNT)	joule	+09 4.20
watt hour	joule	+03 3.60*

ENERGY/AREA TIME

To convert from	to	multiply by
Btu (thermochemical)/foot² second	watt/meter²	+04 1.134 893 1
Btu (thermochemical)/foot² minute	watt/meter²	+02 1.891 488 5
Btu (thermochemical)/foot² hour	watt/meter²	+00 3.152 480 8
Btu (thermochemical)/inch² second	watt/meter²	+06 1.634 246 2
calorie (thermochemical)/cm² minute	watt/meter²	+02 6.973 333 3
erg/centimeter² second	watt/meter²	−03 1.00*
watt/centimeter²	watt/meter²	+04 1.00*

FORCE

To convert from	to	multiply by
dyne	newton	−05 1.00*
kilogram force (kgf)	newton	+00 9.806 65*
kilopond force	newton	+00 9.806 65*
kip	newton	+03 4.448 221 615 260 5*
lbf (pound force, avoirdupois)	newton	+00 4.448 221 615 260 5*
ounce force (avoirdupois)	newton	−01 2.780 138 5
pound force, lbf (avoirdupois)	newton	+00 4.448 221 615 260 5*
poundal	newton	−01 1.382 549 543 76*

LENGTH

To convert from	to	multiply by
angstrom	meter	−10 1.00*
astronomical unit (IAU)	meter	+11 1.496 00
astronomical unit (radio)	meter	+11 1.495 978 9
cable	meter	+02 2.194 56*
caliber	meter	−04 2.54*
chain (surveyor or gunter)	meter	+01 2.011 68*

To convert from	to	multiply by
chain (engineer or ramden)	meter	+01 3.048*
cubit	meter	−01 4.572*
fathom	meter	+00 1.8288*
fermi (femtometer)	meter	−15 1.00*
foot	meter	−01 3.048*
foot (U.S. survey)	meter	+00 1200/3937*
foot (U.S. survey)	meter	−01 3.048 006 096
furlong	meter	+02 2.011 68*
hand	meter	−01 1.016*
inch	meter	−02 2.54*
league (U.K. nautical)	meter	+03 5.559 552*
league (international nautical)	meter	+03 5.556*
league (statute)	meter	+03 4.828 032*
light year	meter	+15 9.460 55
link (engineer or ramden)	meter	−01 3.048*
link (surveyor or gunter)	meter	−01 2.011 68*
meter	wavelengths Kr 86	+06 1.650 763 73*
micron	meter	−06 1.00*
mil	meter	−05 2.54*
mile (U.S. statute)	meter	+03 1.609 344*
mile (U.K. nautical)	meter	+03 1.853 184*
mile (international nautical)	meter	+03 1.852*
mile (U.S. nautical)	meter	+03 1.852*
nautical mile (U.K.)	meter	+03 1.853 184*
nautical mile (international)	meter	+03 1.852*
nautical mile (U.S.)	meter	+03 1.852*
pace	meter	−01 7.62*
parsec (IAU)	meter	+16 3.085 7
perch	meter	+00 5.0292*
pica (printers)	meter	−03 4.217 517 6*
point (printers)	meter	−04 3.514 598*
pole	meter	+00 5.0292*
rod	meter	+00 5.0292*
skein	meter	+02 1.097 28*
span	meter	−01 2.286*
statute mile (U.S.)	meter	+03 1.609 344*
yard	meter	−01 9.144*

MASS

carat (metric)	kilogram	−04 2.00*
gram (avoirdupois)	kilogram	−03 1.771 845 195 312 5*
gram (troy or apothecary)	kilogram	−03 3.887 934 6*
grain	kilogram	−05 6.479 891*
gram	kilogram	−03 1.00*
hundredweight (long)	kilogram	+01 5.080 234 544*
hundredweight (short)	kilogram	+01 4.535 923 7*
kgf second² meter (mass)	kilogram	+00 9.806 65*
kilogram mass	kilogram	+00 1.00*
lbm (pound mass, avoirdupois)	kilogram	−01 4.535 923 7*
ounce mass (avoirdupois)	kilogram	−02 2.834 952 312 5*
ounce mass (troy or apothecary)	kilogram	−02 3.110 347 68*
pennyweight	kilogram	−03 1.555 173 84*
pound mass, lbm (avoirdupois)	kilogram	−01 4.535 923 7*
pound mass (troy or apothecary)	kilogram	−01 3.732 417 216*
scruple (apothecary)	kilogram	−03 1.295 978 2*
slug	kilogram	+01 1.459 390 29
ton (assay)	kilogram	−02 2.916 666 6

To convert from	to	multiply by
ton (long)	kilogram	+03 1.016 046 908 8*
ton (metric)	kilogram	+03 1.00*
ton (short, 2000 pound)	kilogram	+02 9.071 847 4*
tonne	kilogram	+03 1.00*

POWER

To convert from	to	multiply by
Btu (thermochemical)/second	watt	+03 1.054 350 264 488
Btu (thermochemical)/minute	watt	+01 1.757 250 4
calorie (thermochemical)/second	watt	+00 4.184*
calorie (thermochemical)/minute	watt	−02 6.973 333 3
foot lbf/hour	watt	−04 3.766 161 0
foot lbf/minute	watt	−02 2.259 696 6
foot lbf/second	watt	+00 1.355 817 9
horsepower (550 foot lbf/second)	watt	+02 7.456 998 7
horsepower (boiler)	watt	+03 9.809 50
horsepower (electric)	watt	+02 7.46*
horsepower (metric)	watt	+02 7.354 99
horsepower (U.K.)	watt	+02 7.457
horsepower (water)	watt	+02 7.460 43
kilocalorie (thermochemical)/minute	watt	+01 6.973 333 3
kilocalorie (thermochemical)/second	watt	+03 4.184*
watt (international of 1948)	watt	+00 1.000 165

PRESSURE

To convert from	to	multiply by
atmosphere	newton/meter²	+05 1.013 25*
bar	newton/meter²	+05 1.00*
barye	newton/meter²	−01 1.00*
centimeter of mercury (0° C)	newton/meter²	+03 1.333 22
centimeter of water (4° C)	newton/meter²	+01 9.806 38
dyne/centimeter²	newton/meter²	−01 1.00*
foot of water (39.2° F)	newton/meter²	+03 2.988 98
inch of mercury (32° F)	newton/meter²	+03 3.386 389
inch of mercury (60° F)	newton/meter²	+03 3.376 85
inch of water (39.2° F)	newton/meter²	+02 2.490 82
inch of water (60° F)	newton/meter²	+02 2.4884
kgf/centimeter²	newton/meter²	+04 9.806 65*
kgf/meter²	newton/meter²	+00 9.806 65*
lbf/foot²	newton/meter²	+01 4.788 025 8
lbf/inch² (psi)	newton/meter²	+03 6.894 757 2
millibar	newton/meter²	+02 1.00*
millimeter of mercury (0° C)	newton/meter²	+02 1.333 224
pascal	newton/meter²	+00 1.00*
psi (lbf/inch²)	newton/meter²	+03 6.894 757 2
torr (0° C)	newton/meter²	+02 1.333 22

SPEED

To convert from	to	multiply by
foot/hour	meter/second	−05 8.466 666 6
foot/minute	meter/second	−03 5.08*
foot/second	meter/second	−01 3.048*
inch/second	meter/second	−02 2.54*
kilometer/hour	meter/second	−01 2.777 777 8
knot (international)	meter/second	−01 5.144 444 444
mile/hour (U.S. statute)	meter/second	−01 4.4704*
mile/minute (U.S. statute)	meter/second	+01 2.682 24*
mile/second (U.S. statute)	meter/second	+03 1.609 344*

To convert from	to	multiply by

TEMPERATURE

Celsius	kelvin	$t_K = t_C + 273.15$
Fahrenheit	kelvin	$t_K = (5/9)(t_F + 459.67)$
Fahrenheit	Celsius	$t_C = (5/9)(t_F - 32)$
Rankine	kelvin	$t_K = (5/9)t_R$

TIME

day (mean solar)	second (mean solar)	+04 8.64*
day (sidereal)	second (mean solar)	+04 8.616 409 0
hour (mean solar)	second (mean solar)	+03 3.60*
hour (sidereal)	second (mean solar)	+03 3.590 170 4
minute (mean solar)	second (mean solar)	+01 6.00*
minute (sidereal)	second (mean solar)	+01 5.983 617 4
month (mean calendar)	second (mean solar)	+06 2.628*
second (ephemeris)	second	+00 1.000 000 000
second (mean solar)	second (ephemeris)	Consult American Ephemeris and Nautical Almanac
second (sidereal)	second (mean solar)	−01 9.972 695 7
year (calendar)	second (mean solar)	+07 3.1536*
year (sidereal)	second (mean solar)	+07 3.155 815 0
year (tropical)	second (mean solar)	+07 3.155 692 6
year 1900, tropical, Jan., day 0, hour 12	second (ephemeris)	+07 3.155 692 597 47*
year 1900, tropical, Jan., day 0, hour 12	second	+07 3.155 692 597 47

VISCOSITY

centistoke	meter²/second	−06 1.00*
stoke	meter²/second	−04 1.00*
foot²/second	meter²/second	−02 9.290 304*
centipoise	newton second/meter²	−03 1.00*
lbm/foot second	newton second/meter²	+00 1.488 163 9
lbf second/foot²	newton second/meter²	+01 4.788 025 8
poise	newton second/meter²	−01 1.00*
poundal second/foot²	newton second/meter²	+00 1.488 163 9
slug/foot second	newton second/meter²	+01 4.788 025 8
rhe	meter²/newton second	+01 1.00*

VOLUME

acre foot	meter³	+03 1.233 481 837 547 52*
barrel (petroleum, 42 gallons)	meter³	−01 1.589 873
board foot	meter³	−03 2.359 737 216*
bushel (U.S.)	meter³	−02 3.523 907 016 688*
cord	meter³	+00 3.624 556 3
cup	meter³	−04 2.365 882 365*
dram (U.S. fluid)	meter³	−06 3.696 691 195 312 5*
fluid ounce (U.S.)	meter³	−05 2.957 352 956 25*
foot³	meter³	−02 2.831 684 659 2*
gallon (U.K. liquid)	meter³	−03 4.546 087
gallon (U.S. dry)	meter³	−03 4.404 883 770 86*
gallon (U.S. liquid)	meter³	−03 3.785 411 784*
gill (U K.)	meter³	−04 1.420 652
gill (U.S.)	meter³	−04 1.182 941 2
hogshead (U.S.)	meter³	−01 2.384 809 423 92*
inch³	meter³	−05 1.638 706 4*
liter	meter³	−03 1.00*

To convert from	to	multiply by
ounce (U.S. fluid)	meter³	−05 2.957 352 956 25*
peck (U.S.)	meter³	−03 8.809 767 541 72*
pint (U.S. dry)	meter³	−04 5.506 104 713 575*
pint (U.S. liquid)	meter³	−04 4.731 764 73*
quart (U.S. dry)	meter³	−03 1.101 220 942 715*
quart (U.S. liquid)	meter³	−04 9.463 529 5
stere	meter³	+00 1.00*
tablespoon	meter³	−05 1.478 676 478 125*
teaspoon	meter³	−06 4.928 921 593 75*
ton (register)	meter³	+00 2.831 684 659 2*
yard³	meter³	−01 7.645 548 579 84*

INDEX